The Urban Transport Crisis in Europe and North America

John Pucher
Associate Professor of Urban Planning
Rutgers University

and

Christian Lefèvre
Associate Professor of Urban Studies
Institut d'Urbanisme de Paris

First published 1996 by
MACMILLAN PRESS LTD
Houndmills, Basingstoke, Hampshire RG21 6XS
and London
Companies and representatives
throughout the world

ISBN 0–333–62795–4 hardcover
ISBN 0–333–65551–6 paperback

A catalogue record for this book is available
from the British Library.

10 9 8 7 6 5 4 3 2 1
05 04 03 02 01 00 99 98 97 96

Printed and bound in Great Britain by
Antony Rowe Ltd,
Chippenham, Wiltshire

Contents

List of Tables and Figures

Tables

Figures

Preface

This book represents the culmination of the authors' many years of research on urban transport. Pucher began studying American transport policy in 1975, with emphasis on the efficiency and equity impacts of subsidies. Since 1984, his efforts have focused on comparative analysis of transport policies in Europe and North America. Lefèvre's interests have also been in comparative policy analysis but, in contrast to Pucher's economic focus, Lefèvre has concentrated on political and institutional issues. The two approaches complement each other. The book offered the authors an opportunity to update their previous analyses, extend them to more countries, and compare them with each other's different range of experiences. Thus, the authors were able to learn a great deal themselves in the process of writing this book.

The listed order of authorship was chosen at random and has no implication for the relative contributions of each author. Although the idea for the book and its approximate content was joint, each author focused his effort on the subject area of the book closest to his own interests and knowledge. Thus, Pucher wrote the four chapters on the USA, Canada, Germany and Eastern Europe. Lefèvre wrote the four chapters on Italy, France, Great Britain and The Netherlands. Overall, however, the book represents a genuine collaboration, with both authors involved in determining the structure of the book and individual chapters, data assembly and analysis, and policy evaluation.

The authors are deeply indebted to a great number of individuals and institutions for their assistance in obtaining information essential to writing the book. Each author would like to thank in turn those who helped. For the chapter on Germany, Pucher thanks the German Ministry of Transport, the German Association of Transport Firms (especially Mr Wergles), the German Institute of Economic Research (Dr Kunert), and many individual public transport systems for essential data and advice. In addition, he is deeply indebted to the transportation institutes at the University of Muenster and the Technical University of Dresden which hosted Pucher for extended periods to do research on German transport. The Humboldt Foundation of the German Federal Government funded Pucher's research on Germany and Eastern Europe. For the chapter on Eastern Europe, Pucher thanks the Polish, Czech, German and Hungarian ministries of transport for providing extensive assistance. Moreover, colleagues at universities in Dresden (Professors Voigt, Ackermann, Foerschner), Prague (Dr Klofac and Professors Sykora and Cepek), Budapest (Professor Koller) and Warsaw (Dr Rataj, Dr Kocon, Professors Suchorzewski, Lijewski and Weglenski) greatly facilitated the research on Eastern Europe by offering data, analysis, and friendship. The US Department of Transportation and the American Public Transit Association were the most important sources of information for the chapter on the USA. In addition, the US Department of Transportation funded several of Pucher's research projects between 1980 and 1990, which yielded information for the chapter. The Canadian Urban Transit Association, Statistics Canada and Transport Canada provided most of the data for the Canada chapter. Pucher is especially indebted to colleagues in Calgary (Professor Perl), York (Professor Turritin), and Toronto (Dr Gurin, Dr Cormier, Dr Hemily and Dr Dunbar) for providing much information, analysis and advice.

For the chapter on France, Lefèvre thanks his colleague Dr Offner in the LATTS (Laboratoire des Techniques, Territoires et Société) and various institutions such as the French Ministry of Transport, Centre d'Etudes des Transports Urbains (CETUR), Institut National de Recherche sur les Transports et leur Sécurité (INRETS), Groupement des Autorités Responsables des Transports (GART) and Union des Transports Publics (UTP) for their advice and assistance. For the chapter on Italy, Lefèvre thanks Professor Podesta (Politecnico di Milano) and the Centro Studi Sui Sistemi di Trasporto (in Rome) for their comments on the draft version. He would also like to thank those who aided his research on transport in Milan area: Mr Anselmetti (of the regional government), Mr Minotti (of the Milan planning agency), Mr Spatti (of the Regional Railways), and Mr Tarulli (of the Istituto Regionale di Ricerca dello Combardia). For the UK, Lefèvre thanks Professor White (University of Westminster) for his useful suggestions on improving various drafts of the chapter, and also the staff of the Transport and Road Research Laboratory (TRRL) and the Greater Manchester Transport Executive. For the Netherlands, Lefèvre thanks Mr Baanders in the research division of the Dutch Ministry of Transport and Professor Jansen (Institute of Spatial Organization in Delft).

Both authors would like to thank the European Conference of the Ministers of Transport (ECMT) and the International Union of Public Transport (UITP), especially its General Secretary, Pierre Laconte, for their overall assistance throughout the writing of this book. They provided extensive data directly and also facilitated contacts with individual ministries of transport, transport associations and particular public transport systems. Bo Peterson (of AB Storstockholms Lokaltrafik) provided supplemental information on the situation of public transport in Sweden.

JOHN PUCHER CHRISTIAN LEFÈVRE

List of Acronyms

ATM	Azienda Trasporti Municipali
BR	British Rail
CETUR	Centre d'Etudes des Transports Urbains
CNT	Conto Nazionale dei Trasporti
DB	Deutsche Bundesbahn
DoE	Department of the Environment
DOT	Department of Transport
ECMT	European Conference of the Ministers of Transport
EU	European Union
FNM	Ferrovie Nord Milano
FNT	Fondo Nazionale dei Trasporti
FS	Ferrovie dello Stato
GART	Groupement des Autorités Responsables des Transports
GLC	Greater London Council
HOV	High Occupancy Vehicle
INRETS	Institut National de Recherche sur les Transports et leur Sécurité
INSEE	Institut National de la Statistique et des Etudes Economiques
IRF	International Road Federation
ISTEA	Intermodal Surface Transportation and Efficiency Act, 1991
IVHS	Intelligent Vehicle Highway System
LBL	London Bus Ltd
LOTI	1982 Law on domestic transport
LRT	Light Rail Transit
NBC	National Bus Company
NS	Nederlandse Spoorswegen
NTS	National Travel Survey
OECD	Organisation for Economic Cooperation and Development
PDU	Plan de Déplacements Urbains
PTA	Passenger Transport Authority
PTC	Passenger Transport Company
PTE	Passenger Transport Executive
PTU	Périmètre de Transport Urbain
RATP	Régie Autonome des Transports Parisiens
RIF	Région Ile de France
SBG	Scottish Bus Group

TGV	Train à Grande Vitesse
TRRL	Transport and Road Research Laboratory
TSM	Traffic System Management
UITP	International Union of Public Transport
UTP	Union des Transports Publics
VAL	Véhicule Automatique Léger
VDV	Verband Deutscher Verkehrsunternehmen
VT	Versement Transport

1 The Urban Transport Crisis: Introduction

Transport problems have plagued cities from their very beginnings. Even horse drawn chariots caused congestion and safety problems in ancient Rome. Steam locomotives in the nineteenth century were both noisy and polluting, leading to the peripheral locations of railroad stations in most European cities. Tram accidents caused thousands of deaths in both America and Europe during the early twentieth century. Urban streets were filled with trams, omnibuses, pedestrians and horse drawn freight wagons, leading to frequent traffic jams. Transport problems, therefore, are not new, but they have been greatly exacerbated by the automobile, which has caused much more serious and more extensive social and environmental problems than ever before: more noise, air and water pollution, accident and injury, congestion, energy waste, urban sprawl, social segregation and inequity in mobility.

As transport technology improved over the centuries, the speed and comfort of travel increased, but the urban transport problem was never solved. Instead, it became more extensive and more complicated. Mobility has greatly increased, but cities have decentralized so much that accessibility has fallen for much of the population. The spreading out of land use has steadily increased the amount of travel required to reach shopping, recreational, educational and employment locations. Increased travel requirements have largely offset increases in the speed of travel. Thus, as numerous studies of long-run trends have shown, roughly the same amount of time has been spent on daily travel needs in cities even though travel speeds have risen several fold.

The enormous surge in the total amount of travel has caused most of the transport problems in today's cities. Congestion is the most obvious example of too much travel demand compared to transport supply. But most other transport problems are also related to excessive amounts of travel. Mobility, which was once viewed as a goal in itself, has become the key problem. Too much travel means not just congested roads but also excessive noise, air pollution, traffic accidents and energy use. Technological improvements can help mitigate these harmful side effects of urban travel, but they will never solve the problem completely; and, in general, the greater the volume of travel, the more significant the overall magnitude of its social and environmental impacts.

Throughout the world, travel has been increasing rapidly over recent decades, and transport problems have been getting worse. Congestion, for example, now plagues cities in virtually all countries, regardless of their degree of economic and technological development. Similarly, the environmental problems of transport are truly international in scope, afflicting cities in both the poorest and richest of countries.

OBJECTIVES OF THE BOOK

European and North American cities are especially interesting cases for examining the nature and extent of the urban transport crisis. They show that advanced technology and affluence do not solve transport problems. On the contrary, they have generated ever more travel and thus exacerbated the urban transport crisis. Of course, there are large differences among the countries of Europe and North America in their transport systems, travel behaviour, transport problems and policies, and it is precisely that variation that is our focus.

In this book, we examine the urban transport crisis from an international, comparative perspective. Throughout the industrialized world, car ownership and use have grown rapidly over the past few decades. In contrast, public transport use has either fallen or stagnated, so that urban transport has become ever more car oriented. Those trends have caused increasingly severe social, economic and environmental problems. While providing more flexibility, comfort and door-to-door speed for the individual traveller, the automobile causes far more severe external impacts than public transport. Thus, increased car ownership and use are directly related to increased pollution and congestion in European and North American cities.

The general trend towards more car ownership and more car-dependent cities is similar in most countries, but the current levels of ownership and use vary greatly. Likewise, the trend towards decentralization of cities can be found throughout Europe and North America, but the current extent of suburbanization and sprawl in urban development varies enormously from one country to another. The variations in travel behaviour and land use patterns have important implications for the overall severity of each country's urban transport crisis and which problems are most serious.

The objective of the book, of course, is not just to describe differences in transport systems, travel behaviour, and transport problems. Perhaps most importantly, we seek to identify policies that would help solve or at least mitigate urban transport problems. An international, comparative approach is crucial for achieving that goal, since different countries have adopted quite different transport policies. By examining the successes and failures of policies in different countries, we hope to help countries learn from each other's experiences.

ORGANIZATION OF THE BOOK

Our analysis is divided into three main parts: the introduction and overview (Chapters 1 and 2); in-depth case studies of eight countries (Chapters 3 to 10); and the summary, comparative evaluation and policy recommendations (Chapter 11).

Chapter 2 provides an overview of urban transport in Europe and North America, including comparisons of trends in car ownership, travel behaviour, road supply, public transport services and urban land use patterns. We discuss a range of transport problems, their relative importance, and how they vary by country and over time. Chapter 2 concludes with a comparison of public policies towards urban transport in the various countries, considering dimensions such as subsidy policy, transport pricing and taxation, land use controls, restraints on car use, privatization and deregulation, and laws mandating technological solutions.

Chapters 3–10 constitute the bulk of the book. Those chapters provide an in-depth analysis of the urban transport situation in each of eight countries (or groups of countries). The exact content varies somewhat from country to country but follows basically the same lines. A brief introduction highlights the distinguishing characteristics of transport in each country. Then we examine developments in urban spatial structure, trends in transport supply and demand, the nature and extent of each country's urban transport crisis, and the public policy responses to urban transport problems. Each of the country chapters concludes with an overall assessment of problems, policies and likely future developments.

Germany (Chapter 3) was chosen for detailed analysis because it is the largest country in the EU. It also has the third highest population density in Europe, with severe problems of air pollution, safety and congestion. Moreover, it has led the world in regional multi-modal coordination of public transport. Although it has one of the world's highest levels of car ownership, Germany has made massive investments in its public transport system and has done much to restrict automobile use through traffic calming, pedestrian zones and parking supply limitations. Finally, it has one of the strictest land use policies in Europe, helping to discourage the extensive low density sprawl found in many other countries.

France (Chapter 4) is one of the other key countries in the EU, especially in terms of its influence in setting overall policies for the Union. It is also interesting due to the huge investment it made in a wide range of public transport systems during the decades of the 1970s and 1980s, and its unique method of financing public transport subsidies through employer payroll taxes. France has the longest history of widespread private management and operation of public transport systems, which is an interesting contrast to most other countries and also to its own high level of public subsidies.

The Netherlands (Chapter 5) is unique among the countries of the developed world in the extraordinary importance of non-motorized modes of urban transport. Bicycling accounts for a much higher proportion of travel than anywhere else, and public policies have reinforced their historical importance by continuously improving cycling networks and pedestrian facilities. Traffic calming and pedestrian zones had their origins in the Netherlands. Yet another reason for devoting an entire chapter to the Netherlands is the superb integration of public transport services and the very centralized planning, operation and financing of public transport (the most centralized in all of Europe).

Italy (Chapter 6) is a fascinating case due to the nearly chaotic urban transport situation there. Somehow transport systems seem to function and people do manage to get around in spite of the lack of any coherent transport policies. Neither road policies nor public transport systems are coordinated; they seem to evolve on their own, without any sort of meaningful plans. Of course, that very lack of planning and coordinated transport policy has caused severe congestion and pollution problems in Italian cities.

The UK (Chapter 7) presents a dramatic case of how important public policy can be in determining the fate of urban transport. Through the decentralization, privatization and fiscal austerity policies of Thatcher, public transport systems in England have been decimated. Service changes and fare increases have led to massive ridership losses and accelerated the modal shift to the automobile. That very shift from public transport to private car makes the UK an interesting case for analysis.

Eastern Europe (Chapter 8) provides yet another example of sudden shifts from public to private transport. Prior to the various revolutions that overthrew the Communist dictatorships in East Germany, Poland, Czechoslovakia and Hungary, urban transport was almost completely dominated by public transport. In the years since 1989, however, public transport has lost many riders, and car ownership and use have skyrocketed, causing extremely serious problems of congestion, pollution and safety. Unlike the case of Great Britain, however, where transport policy was deliberately changed after extensive analysis, the changes in Eastern Europe resulted from fundamental changes in economic and political structures.

Crossing the Atlantic, we then turn to the two most important North American countries. Canada (Chapter 9) plays a special role as a sort of bridge between Europe and the USA. On the one hand, its cities are far more compact and less suburbanized than American cities, but less dense and more suburbanized than most European cities. Similarly, Canadian cities offer much more extensive and higher quality public transport services than American cities, and public transport accounts for about three to four times as high a percentage of urban travel as in the USA. Nevertheless, walking and bicycling are as insignificant in Canada as in the USA, and the automobile accounts for almost twice as high a percentage of total urban travel as in Europe. Canada is also interesting due to the almost total decentralization of urban transport planning, operations and finance. The federal government in Ottawa plays virtually no role in urban transport.

Finally, we examine the unique case of the USA (Chapter 10), where the car has dominated urban transport longer and more completely than anywhere else in the world. Except for a handful of cities, public transport use is so low that it is insignificant and certainly not considered a feasible alternative. Similarly, metropolitan areas in the USA are already the most fragmented, decentralized, suburbanized and lowest density in the world. The extensive, low density urban sprawl makes the automobile a virtual necessity for any travel purpose. Even massive subsidies to public transport over the past 20 years have failed to halt the increase in car use. The USA represents the extreme case where policies have fostered ever more suburbanization and vehicle use, generating higher and higher levels of mobility, but not necessarily greater accessibility. The long history of automobile dependence in the USA has also meant extensive experience with the many social and environmental problems of excessive car use. That experience should be interesting and useful for European countries now grappling with the same problems.

The third section (Chapter 11) of the book summarizes the differences among countries and trends over time in urban transport. Most of the final chapter, however, is devoted to evaluating the effectiveness of alternative policies for dealing with urban transport problems. For that evaluation, we draw in particular on the analysis in the eight case studies of individual countries, but we also examine evidence from other countries and from other sources in the literature. The emphasis here is on what countries can learn from each other's experiences with urban transport problems and policies. The chapter concludes with a set of policy recommendations to improve transport policies, and a discussion of likely future developments in urban transport.

2 Overview of Urban Transport Systems and Land Use Patterns in Europe and North America

The countries of Europe and North America all seem to be headed in the same direction: more car ownership and use. As shown in this chapter, however, current levels of car ownership and especially use vary greatly from one country to another. For example, the car accounts for almost twice as high a proportion of urban travel in North America as in Western Europe, and four times as high as in Eastern Europe. Conversely, walking and bicycling account for roughly three to five times as high a proportion of urban travel in Europe as in either the USA or Canada. Public transport serves four to six times as high a percentage of urban trips in Canada and Europe as in the USA. Although the world-wide trend towards more automobile ownership and use has indeed produced some convergence, differences in urban transport systems and travel behaviour remain significant.

Similarly, cities throughout Europe and North America have been decentralizing in recent decades. Both population and employment have been moving to the suburbs. Nevertheless, the suburban development around American cities is vastly different from new development at the fringes of European cities. Not only are European suburbs less extensive than American suburbs, but they are also much denser and less dependent on the car for mobility needs. Thus, spatial structure remains a crucial distinguishing characteristic between European and American cities.

This chapter examines recent trends in urban transport systems, travel behaviour and land use patterns, documenting similarities between Europe and North America in the general trend towards increased car use and suburbanization but emphasizing the significant differences that persist among countries. We also explore the range of transport problems and how they vary in importance from country to country: air pollution, congestion, energy, equity, financing, safety and urban sprawl. In some cities, transport problems have reached truly crisis proportions, while in others they are simply annoying inconveniences. Thus, the overall importance of the urban transport problem, as well as the relative importance of different kinds of transport problems, can vary from country to country and even from city to city within the same country.

We conclude by describing the wide range of public policies towards urban transport, how they vary among countries, and how they have changed in recent decades. For example, we examine the apparent shift from supply-side strategies aimed at increasing the capacity of transport systems to demand-side strategies aimed at reducing or redirecting trip volumes, patterns and modal choice by altering travel behaviour. We also examine the trends towards greater decentralization, deregulation and privatization in urban transport policies found in most countries of Europe and North America. In this chapter, we limit ourselves mainly to describing government policies. We postpone our summary evaluation of those policies until Chapter 11, where we compare the relative advantages and disadvantages of alternative transport policies, based largely on the detailed analysis of the eight individual country chapters.

PROBLEMS OF DATA RELIABILITY AND COMPARABILITY

Before examining urban transport trends for various countries, we would like to emphasize that there are certain inevitable problems of data comparability. Although every effort has been made to use statistics that are, in fact, comparable across countries, different methods of collecting data and calculating the various indices may cause distortions. Different countries may have somewhat different definitions and statistical interpretations of the same concepts.

Even the most basic of statistics, such as vehicle registrations per capita, can be measured in different ways and have different meanings. Automobiles can take on quite different manifestations, ranging from mini-autos in some European countries to multi-purpose vans and mini-trucks used for passenger transport in the USA. Different sized vehicles obviously have different implications for congestion, pollution and energy use, but available statistics do not allow us to distinguish the growing stock by size and type for all countries.

There is particularly wide variation in the terminology used to describe public transport modes. In the USA and Canada, for example, urban public transport in general is called 'mass transit' or 'public transit', or simply 'transit'. Trams are called 'streetcars', although modern, more advanced tramway systems are called 'light rail transit' (LRT). Metros and underground railways are called 'subways', 'elevateds', 'rapid transit', or 'heavy rail transit'. Suburban rail services are called 'commuter rail'. For consistency and brevity, we will generally use the British terminology in this book.

One of the most basic data problems is the different categorization of transport into urban versus non-urban (rural or interurban) transport. Some countries distinguish only between local and non-local transport, or between short-distance and long-distance transport. For those countries without separate urban transport statistics, we have treated local or short-distance transport as urban transport since there was no other alternative.

Trip definitions also vary. Some countries collect travel data on the basis of comprehensive origin-to-destination trips, reporting only the main type of transport used. Other countries rely on so-called 'unlinked' trip data, considering every transfer from one mode to another an additional trip and recording it separately.

Other problems of data definition and comparability will be noted as they arise. We would like to warn the reader at the outset, however, that some degree of inconsistency is unavoidable. We will clearly indicate those instances where data problems seem to be most serious.

Where there are alternatives, we have chosen to use data from widely recognized agencies such as the World Bank, the Organization for Economic Cooperation and Development (OECD), the UN, the UITP and the International Road Federation (IRF). In many instances, however, the only source was the transport ministry or public transport association in each of the various countries, and there was no way to countercheck the accuracy of those data.

BASIC DEMOGRAPHIC TRENDS

Table 2.1 contains basic demographic information for various OECD countries, including the most important countries of North America and Europe. Japan is included as a basis for comparison.

Overall, population has been growing roughly twice as fast in North America as in Europe. During the two decades from 1970 to 1990, growth exceeded 22 per cent in both Canada and the USA. In Europe, growth ranged from less than 1 per cent in Hungary to 4 per cent in Italy, 11 per cent in France, and 15–18 per cent in the Netherlands, Greece, Poland and Spain.

For many decades, the degree of urbanization has been increasing throughout North America and Europe. By 1990, the vast majority of the population on both continents was already living in cities. Nevertheless, the degree of urbanization varies considerably among individual countries, ranging from only 34 per cent in Portugal to about 75 per cent in Canada, the USA, Germany and France to 96 per cent in Belgium.

There are even larger differences among countries in population density. Population density is much lower in North America than Europe,

Table 2.1 Population, population change, density and rate of urbanization in selected countries, 1970–92

	Population (in thousands)			*% change*	*Density (persons per sq. km)*			*% urban*		
Country	*1970*	*1980*	*1990*	*1970–90*	*1970*	*1980*	*1990*	*1970*	*1980*	*1990*
Austria	7 398	7 553	7 718	4	89	91	92	52	55	58
Belgium	9 691	9 863	9 987	3	316	323	328	94	95	96
Canada	21 297	24 091	26 520	24	2	2	3	76	76	77
Czechoslovakia	14 443	15 280	15 675	8	114	121	124	55	67	77
Denmark	4 943	5 124	5 146	4	115	119	121	80	84	85
Finland	4 679	4 787	4 998	7	14	14	15	50	60	60
France	50 900	54 000	56 600	11	94	99	104	71	73	73
Germany	61 500	61 658	64 485	5	245	248	257	80	83	85
Greece	8 729	9 700	10 100	16	67	73	77	52	58	62
Hungary	10 347	10 713	10 375	0	113	117	115	48	57	64
Italy	54 682	57 100	56 746	4	179	187	194	64	67	69
Japan	103 720	118 010	123 693	19	282	320	336	71	76	77
Netherlands	13 046	14 200	15 010	15	313	339	370	86	88	89
Norway	3 891	4 092	4 250	9	12	13	13	65	70	75
Poland	32 605	35 735	38 143	17	106	116	122	52	58	62
Portugal	8 124	9 867	9 950	22	94	107	107	26	29	34
Spain	34 032	37 430	39 987	17	67	75	78	66	73	78
Sweden	8 091	8 317	8 589	6	18	18	19	81	83	84
Switzerland	6 224	6 329	6 637	7	152	155	166	54	57	61
UK	55 780	56 341	57 486	3	229	231	236	88	89	89
USA	204 766	228 830	250 878	23	22	25	27	74	74	75

Sources: Eurostat (1993) and UN (1994).

with by far the lowest average density in Canada (only 3 inhabitants per sq. km). Even the USA (at 27 persons per sq. km) is less dense than all European countries except Norway, Finland and Sweden. The Netherlands, Belgium and Germany are the most densely populated European countries (over 250 persons per sq. km), with France, Portugal and Austria near the European average of about 100.

Finally, Table 2.2 presents information on the range of per capita incomes (per capita GDP) and car ownership levels in the various OECD countries. Trends in ownership and use will be discussed later in this chapter in greater detail, but it is noteworthy that there is a strong correlation between income and car ownership. The much higher price of car ownership and use in Western Europe for many decades probably explains why those countries have somewhat lower levels of ownership per capita

Table 2.2 Per capita income and car ownership in selected countries, 1992 (US$)

Country	*Per capita income*[*]	*Car ownership per 1000 population*
Austria	21 139	410
Belgium	20 007	401
Canada	21 248	486
Czechoslovakia	2 129	219
Denmark	25 298	310
Finland	24 091	383
France	23 170	420
(West) Germany	24 552	492
Greece	6 954	178
Ireland	12 668	235
Italy	20 297	490
Japan (1990)	24 765	313
Netherlands	21 218	374
Norway (1990)	24 888	380
Poland (1988)	1 818	138
Portugal	6 964	182
Romania	1 241	70
Spain	12 622	335
Sweden	27 600	419
Switzerland	34 168	451
UK	14 483	375
USA	22 278	600

[*]GDP per capita taken as income per capita, 1992 (or latest available year).
Sources: IMF, *Government Finance Statistics Year Book* (1993); IRF, *World Road Statistics* (1993).

than might be expected from their high per capita incomes, at least relative to the USA.

TRENDS IN TRAVEL BEHAVIOUR

Car ownership

Surely the most dramatic change in passenger transport over recent decades has been the shift towards ever more car ownership and use throughout the world. That is confirmed by Figure 2.1 and Table 2.3.

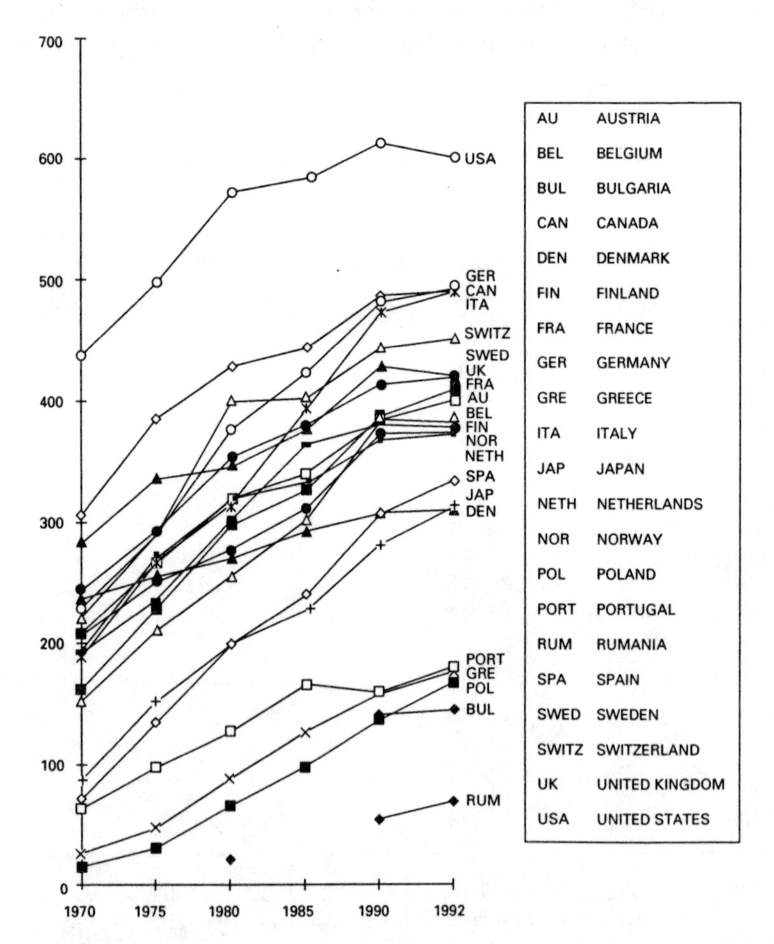

Figure 2.1 Car ownership rate in selected countries, 1970–92 (number of cars per thousand persons)

Source: IRF, *World Road Statistics* (1993).

Table 2.3 Vehicle km of car use per capita, 1970–92

	1970	*1975*	*1980*	*1985*	*1992*
Austria			3 430	3 637	4 147***
Belgium	2 231	2 804	4 143	4 270	5 336**
Canada	6 395	7 352	8 530		8 746
Czechoslovakia					1 642
Denmark	3 544		4 254	4 735	5 790
Finland	2 543	2 926	4 429	5 301	7 131
France	2 615	3 629	4 463	4 739	5 824
Germany	3 366	4 269	4 830	5 136	6 228**
Greece	261	784	*1 064	930	
Hungary	314	776			1 653**
Italy	2 240	2 851	3 338	3 733	5 438
Japan	1 162	1 573	2 046	2 405	3 108
Netherlands	3 329	3 801	4 021	4 474	5 270
Norway	2 230	3 123	3 617*		5 373
Poland	127	284	573	541	1 905
Portugal			1 945*	2 327	2 908
Spain	749	1 132	1 410	1 458	2 003
Sweden				6 329	6 713***
Switzerland	3 177				5 184
UK	2 976	3 539	4 108	4 165	6 000
USA	7 001	7 712	7 819	8 472	9 728

*1979; ** 1990; *** 1991.
Sources: IRF; Eurostat; DOT, *Transport Statistics Great Britain.*

Without exception, car ownership grew from 1970 to 1992 in all the countries of Europe and North America, but at considerably different rates. Those countries with the lowest rates of ownership in 1970 experienced the most rapid increases, while those with the highest rates in 1970 experienced the slowest growth. Thus, car ownership increased fastest in Eastern and Southern European countries, which had fewer than 100 cars per 1000 population in 1970. Northwestern European countries, which had 150–250 cars per 1000 population in 1970, had much slower rates of growth. Canada and the USA, with the highest starting levels of ownership, had the smallest increases. That pattern of differential growth rates is narrowing the gap between the most and least car-oriented countries.

Particularly striking is the actual decline in ownership rates between 1990 and 1992 in the USA, Finland, Denmark, Norway and Sweden. That may be a temporary reversal of long-term trends due to high unemployment during those years. Nevertheless, the slow-down in growth of ownership, and in particular the downturn in a few countries, suggests a certain saturation level, which differs from country to country depending on specific circumstances.

Car use

Figures on vehicle kilometres of use are more difficult to obtain than for car ownership. Indeed, many countries do not report this statistic at all. Moreover, the available information is of questionable reliability and comparability. Unlike vehicle registrations, the amount of vehicle use cannot be directly measured, but must be indirectly estimated on the basis of partial survey data and assumptions about trip frequencies and lengths. Different countries employ somewhat different procedures. Notwithstanding those reservations, Table 2.3 clearly indicates a strong trend towards increased car use in all countries. Similar to the trends for car ownership, use has also been increasing most in those countries that had the least use in 1970.

Unfortunately, it is virtually impossible to obtain consistent, comparable figures for car ownership and use exclusively for urban areas in each country. Only a few countries keep separate statistics for urban as opposed to rural and interurban use and, even for those countries, there are differences in the specific criteria used to isolate urban travel. Nevertheless, the figures on overall car use also reflect a rapid increase in urban use, since urban travel represents a considerable portion of total travel and because urban and non-urban automobile use are affected by similar economic, social and technological factors.

That inference is confirmed by the results of a 1993 study commissioned by the OECD. It included a survey of travel behaviour and transport problems in 132 cities, including 93 cities in 15 European countries and 14 cities in North America (Sharman and Dasgupta, 1993). The European cities reported increases in car traffic ranging from 30 to 35 per cent per decade from 1970 to 1990. Increases in vehicle kilometres of urban travel were even larger, however, in the five US cities included in the survey, ranging from 20 to 80 per cent per decade, with an average increase of about 45 per cent. Underlying those increases in urban car use, all the European and American cities in the OECD survey reported growth in car ownership, although there remain large differences in the absolute levels of ownership per capita. For example, the number of cars per inhabitant in 1990 ranged from 0.5 to 0.7 in American cities, from 0.2 to 0.5 in Western European cities, and from 0.1 to 0.3 in Eastern European cities.

Thus, car ownership and use are increasing in virtually all cities of Europe and the USA, but American cities remain much more car oriented than European cities.

Modal split

The dramatic automobile dependence of American cities is evident in the modal split statistics collected by the 1993 OECD study. The automobile's share of total motorized trips was over 95 per cent in the American cities surveyed by OECD, compared to a range of 40–90 per cent for Western European cities, and a range of 20–50 per cent for Eastern European cities (Sharman and Dasgupta, 1993). Because bicycling and walking are far more important transport modes in European cities, modal split differences between the USA and Europe are even more striking when automobile use is calculated as a percentage of total urban travel and not just motorized travel. In the European cities surveyed, bicycling and walking accounted for 15–50 per cent of total trips in 1990, with an average of about 30 per cent. In sharp contrast, walking and bicycling accounted for less than 5 per cent of total travel in the American cities surveyed. Consequently, the automobile's modal split share of total travel in the USA is almost twice as high as in European cities.

Table 2.4 presents an alternative source of information on modal split: nation-wide aggregate distributions for 12 countries. These modal split distributions are based on travel surveys undertaken by ministries of transport in each country and represent averages for urban areas as a whole. They show that public transport accounts for roughly 10–20 per cent of total urban travel in Western Europe compared to only 3 per cent in the USA. Moreover, the proportion of urban travel by walking and bicycling is over three times higher in Europe than in the USA (34 per cent on average in Europe as against only 10 per cent in the USA).

Of course, there are large differences among cities even within the same countries. Indeed, as shown in Table 2.5, variation within countries can be as great as among countries.

Among the 20 largest American cities, for example, the modal split share of public transport ranges from 28 per cent in in the Greater New York Metropolitan Area to only 2 per cent in Detroit, Phoenix and Dallas. In smaller cities, the public transport share of total travel rarely exceeds 1 percent. Similarly, the modal split share of public transport in Canadian cities ranges from 34 per cent in Montreal and 27 per cent in Toronto to only 5 per cent in smaller cities such as St Catharines. Variation by city size can also be found among European cities, but it is not as extreme.

Table 2.4 Modal split (as percentage of total trips) in urban areas, 1990 (or latest available year)

<table>
<tr><th>Country</th><th>Car</th><th>Public transport</th><th>Two wheelers</th><th>Walking</th><th>Other</th></tr>
<tr><td>Austria</td><td>39</td><td>13</td><td>9</td><td>31</td><td>8</td></tr>
<tr><td>Canada</td><td>74</td><td>14</td><td>1</td><td>10</td><td>1</td></tr>
<tr><td>Denmark</td><td>42</td><td>14</td><td>20</td><td>21</td><td>3</td></tr>
<tr><td>France*</td><td>54</td><td>12</td><td>4</td><td>30</td><td>0</td></tr>
<tr><td>Germany</td><td>52</td><td>11</td><td>10</td><td>27</td><td>0</td></tr>
<tr><td>Italy</td><td>25</td><td>21</td><td colspan="2">54</td><td></td></tr>
<tr><td>Netherlands</td><td>44</td><td>8</td><td>27</td><td>19</td><td>1</td></tr>
<tr><td>Norway**</td><td>68</td><td>7</td><td colspan="2">25</td><td></td></tr>
<tr><td>Sweden</td><td>36</td><td>11</td><td>10</td><td>39</td><td>4</td></tr>
<tr><td>Switzerland</td><td>38</td><td>20</td><td>10</td><td>29</td><td>3</td></tr>
<tr><td>UK***</td><td>62</td><td>14</td><td>8</td><td>12</td><td>4</td></tr>
<tr><td>USA</td><td>84</td><td>3</td><td>1</td><td>9</td><td>2</td></tr>
</table>

*Household surveys of major urban areas.
**For city regions, people aged 18–69.
***England and Wales.
Sources: Primarily from Ministries of Transport for each individual country.

Public transport accounts for over a quarter of travel in large cities such as Paris, Rome and Berlin, but even in smaller cities, it serves 5–10 per cent of trips. Within comparable city size categories, public transport accounts for a much higher proportion of total travel in Europe and Canada than in the USA.

Table 2.6 also highlights the differential importance of public transport among cities within the same country. It reports the average number of public transport trips per inhabitant per year in each of 29 cities. As expected, the lowest values are in the USA (17 in both San Diego and Detroit), and the highest values are in Europe (470 in Zurich and 397 in Vienna).

The OECD study reveals a marked increase in the automobile's share of total urban travel in most European cities from 1970 to 1990, and it also forecasts considerable growth in car ownership, use and modal split by the year 2000. It should be noted, however, that a few European cities have succeeded in reducing vehicle use since 1980 and increasing the modal split shares of public transport, bicycling and walking. For example, Freiburg and Munich (in Germany), Basle and Zurich (in Switzerland) and Vienna (Austria) report falling modal split shares in

Table 2.5 Modal split in selected urban areas (as percentage of total trips), latest available year

Urban area, country, date	*Car*	*Public transport*	*Two wheelers*	*Walking*
New York (USA) *1990	62	28	0	7
Los Angeles (USA) *1990	88	5	1	3
Dallas (USA) *1990	93	2	0	2
Atlanta (USA) *1990	91	5	0	2
Detroit (USA) *1990	93	2	2	0
Boston (USA) *1990	81	11	0	6
Phoenix (USA) *1990	90	2	1	3
Washington (USA) *1990	79	14	0	4
Paris (F) 1991	66	30	4	
Grenoble (F) 1992	54	14	4	27
Bordeaux (F) 1990	64	10	6	20
Toulouse (F) 1990	63	10	6	20
London (UK) 1992	48	17	2	33
Leeds (UK) *1991	60	25	2	13
Sheffield (UK) *1991	58	29	2	11
Bristol (UK) *1991	66	14	6	14
Amsterdam (NL) 1990	38	15	21	26
Utrecht (NL) 1990	40	9	28	22
Munich (D) 1992	36	25	15	24
Wuppertal (D) 1992	54	18	1	27
Freiburg (D) 1991	35	20	20	25
Dresden (D) 1990	43	21	8	28
Genoa (I) 1990	24	48	28	
Rome (I) 1990	51	32	17	
Bergen (N) 1990	57	11	32	
Trondheim (N) 1990	60	10	30	
Uppsala (S) 1990	49	20	31	
Ottawa (CA) 1990	77	13	10	
Helsinki (SF) 1990	46	31	23	
Lausanne (CH) **1990	57	41	2	
Geneva (CH) **1990	53	44	3	
Zurich (CH) **1990	45	54	1	
Prague (CZ) 1990	10	59	3	28

*Data concern only the journey to work.
**Commuting trips of people working in a municipality different from that of their residence. This selection obviously increases the importance of motorized modes. As a comparison, for people living and working in the same municipality, taking Lausanne area as an example in 1990, the modal split is as follows: car, 27%; public transport, 43%; 2 wheelers and walking, 30% (28% walk and 2% 2-wheelers).
Sources: For Genoa down to Helsinki inclusive, figures are elaborated from Sharman and Dasgupta (1993). For British cities, DOT, *Transport Statistics for London* (1993); Dasgupta (1994). For French cities, corresponding household surveys. Swiss cities: LITRA (1994). Dutch cities: Ministry of Transport (1993).

Table 2.6 Public transport trips per inhabitant per year in selected urban areas (1992 or latest available year)

Urban area	*Population in millions*	*Public transport trips/inhabitant*
Montreal (CA)	3.1	195
Toronto (CA)	3.9	186
Vancouver (CA)	1.6	93
Ottawa (CA)	0.9	133
Atlanta (US)	3.1	64
Detroit (US)	4.3	17
Houston (US)	3.4	29
Philadelphia (US)	4.9	78
San Diego (US)	2.5	17
London (UK)	6.7	290
Manchester (UK)	2.6	131
Tyne and Wear (UK)	1.1	340
West Midlands (UK)	2.6	170
Barcelona (E)	2.9	190
Bologna (I)	0.5	278
Milan (I)	3.2	350
Vienna (A)	1.5	397
Amsterdam (NL)	0.7	320
Stockholm (S)	1.5	288
Dusseldorf (D)	1.1	160
Hanover (D)	0.5	230
Cologne (D)	1.1	150
Stuttgart (D)	0.6	250
Bordeaux (F)	0.7	96
Grenoble (F)	0.4	136
Lille (F)	1.0	100
Lyon (F)	1.2	186
Paris (F)	7.2	330
Zurich (CH)	0.6	470

Sources: National public transport associations; UTP.

recent years thanks to vigorous policies aimed at making private car use more difficult and more expensive (Pucher and Clorer, 1992). At the same time, they expanded public transport services, offered lower public transport fares, gave traffic priority to public transport and greatly enhanced facilities for pedestrians and cyclists (car-free zones and bikeways). Thus, it is not inevitable that automobile use must increase at the expense of other modes of travel.

Moreover, there are important variations among European countries in urban transportation developments. Great Britain appears to be much further down the road to the Americanization of its urban transport and land use patterns than any country on the Continent. The shift away from public transport in England and Wales has been dramatic over the past two decades. The proportion of total journeys by all public transport modes fell from 33 per cent in 1971 to only 14 per cent in 1991. Conversely, the automobile's share of travel rose from 37 per cent to 62 per cent. Even walking and bicycling declined relative to car use, their combined modal split share falling from 30 per cent in 1971 to only 23 per cent in 1991 (Dasgupta, 1994). No other Western European country has experienced such a rapid shift towards the automobile.

Policy impacts

As described later in this chapter, public policy differences explain much of the variation within Europe in urban transport trends, just as they explain much of the difference between Europe and the USA. For example, the sharp decline in public transport use in Britain is partly the result of subsidy reductions and fare increases. During the same period, other European countries such as France, the Netherlands and Germany were devoting massive increases in subsidy funds to expanding and improving public transport systems and reducing fares. The result was increased public transport use on the Continent and a slow-down in the shift towards the automobile. In France, for example, total public transport use rose by 58 per cent from 1975 to 1990, a sharp contrast to the 26 per cent decrease in public transport use in Britain over the same period (British Department of Transport, 1993(b); UTP, 1993).

Public transport use

Figure 2.2 shows trends in public transport use for 14 countries in Europe and North America. It is noteworthy that Britain is the only country that experienced a steady decline in public transport use during the entire period, with 1992 ridership only 80 per cent of its 1980 level, and only two-thirds of its 1975 level. All other reporting countries experienced increased use during the decade of the 1970s.

Since 1980, the record has been mixed. Some countries, such as Austria, Czechoslovakia and the Netherlands, saw continued growth in public transport use until 1992. East Germany, Hungary and Poland had increased use until 1985 (in fact, until the overthrow of Communism in

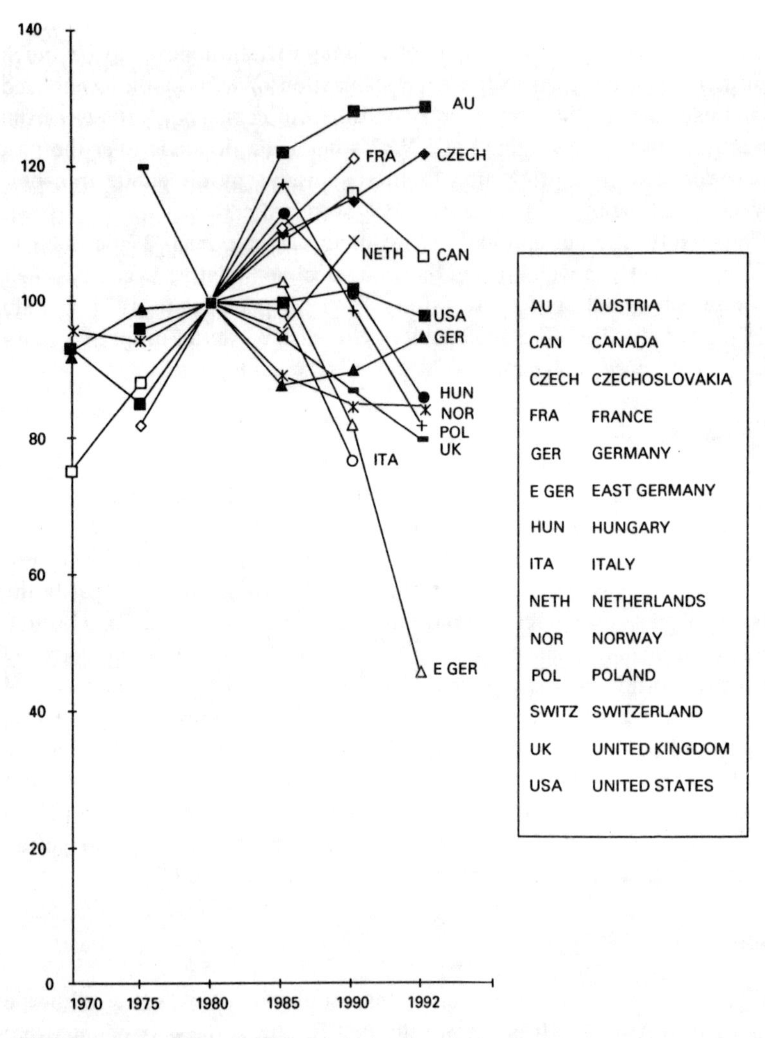

Figure 2.2 Change in public transport ridership in selected countries, 1970–92 (100 in 1980)
Source: National public transport associations.

1988 and 1989), but since then have suffered dramatic declines. Italy's public transport systems lost almost a quarter of their passengers between 1980 and 1990. Public transport use changed only slightly in the USA, whereas in Canada there was a substantial increase until 1990 and then a considerable fall.

Trends in public transport use have, therefore, varied by country and over time. But the modal split share of total urban travel served by public transport has fallen in virtually all countries. Even in countries where total public transport ridership has risen, private vehicle use has risen faster. Although the relative importance of public transport has thus been falling everywhere, European countries still have much higher public transport modal splits than American cities: roughly four to five times as high. Overall trends in urban travel are generally headed in the same directions in the USA and Europe, yet there remain enormous differences in urban transport systems and travel behaviour that are likely to persist. It is virtually inconceivable that European cities will ever become as totally dependent on the automobile as their American counterparts.

TRENDS IN URBAN LAND USE PATTERNS

The main reason for increased car ownership and use throughout the world is income growth, which has made such use more affordable. In addition, various advances in communications and production technologies have encouraged the decentralization of cities. Lower density urban development has both encouraged and necessitated more automobile use, while making public transport, cycling and walking less and less feasible travel options.

The extreme decentralization of population and economic activity within American metropolitan areas has been well documented in hundreds of articles and books. The low density suburban sprawl that surrounds every American city has become a veritable trademark of the American way of life. The OECD study provides evidence that virtually all European and Canadian cities are also decentralizing, although not nearly to the same extent as American cities (Sharman and Dasgupta, 1993).

Not only does a larger proportion of the urban population live in a central city in Europe, but the population density of European suburbs is much higher than in the USA. Newman and Kenworthy (1989) estimate that 42 per cent of the total population in Europe's largest metropolitan areas lives in the central city, compared to only 26 per cent of total population in the 10 largest American metropolitan areas. Moreover, population densities in European suburbs are four times higher than in American suburbs. That suburban density differential is even greater than the central city density differential: population densities in European central cities are twice as high as in American central cities. Similarly, Canadian cities are less suburbanized than American cities, and whatever suburban development has taken place is much denser and more compact than American suburbs (Cervero, 1986a; Pucher, 1994a).

Finally, suburbanization of firms appears to be occuring far more slowly in Europe than in the USA. Newman and Kenworthy (1989) find that 20 per cent of all employment in the largest European metropolitan areas is still in the central business district, compared to an average of only 12 per cent in America's largest cities. As was the case with suburban residential densities, suburban employment densities are also much higher in Europe, more than three times as high as in the USA.

Of course, all the figures referred to in the preceding paragraphs are averages that ignore the considerable differences among different countries within Europe, and even more so the variations within individual countries. For example, urban areas in the northeastern USA and eastern Canada are far less decentralized than urban areas in the southern or western USA and western Canada. The central city's proportion of total metropolitan population and employment in older cities such as Boston, Philadelphia, New York, Montreal and Toronto is substantially higher than in younger cities such as Atlanta, Houston, Denver, Phoenix, Edmonton and Calgary.

In Europe, the differences appear to be along north–south lines rather than east–west lines. Thus, northern European countries such as France, Sweden, England and the Netherlands have been experiencing more suburbanization of their cities than southern European countries such as Portugal, Spain, Greece or Italy. Indeed, France and England are showing increasing signs of 'ex-urban' development (the rapid growth of distant towns and rural areas not contiguous to the primary, built-up metropolitan area). Such ex-urban development is already far advanced in the USA, but it is notable that it is now beginning even in certain European countries. Within countries there are also differences. Thus, northern Italy's cities have seen more suburbanization than southern Italy's cities. Not only has the extent of suburbanization varied, but the type of suburbanization has also varied. In major metropolitan areas of the Netherlands and Germany, for example, urban decentralization has taken the form of massive conurbations such as the Randstad (Amsterdam-Rotterdam) and the Ruhr Region (Essen-Duisburg-Bochum) extending over several thousand square kilometres. In contrast, decentralization in Spain, Portugal and the Scandinavian countries manifests itself in the more traditional form of a single central city with its radially extended, dependent suburbs.

Notwithstanding such variations, the most recent evidence does indeed confirm that European cities are decentralizing, both in terms of population and employment, but that decentralization is coming much later than in the USA and remains much less extensive. Moreover, whatever suburbanization is occurring in Europe is at a much higher density than in the USA. As in the case of increased car ownership and use, the direction of urban development trends is the same in Europe and the USA, but differences in the amount and density of suburbanization remain significant.

There is one more important difference that distinguishes American cities from European and Canadian cities: their extreme degree of socioeconomic segregation. Of course, throughout the world one finds cities whose various quarters are differentiated by income, class, ethnic background and other social characteristics. The spatial separation of different racial, ethnic and economic classes is particularly striking in the USA, however, with suburbs reserved primarily for affluent Whites, and the inner cities for Blacks, Hispanics and the poor of every race and ethnic group. The middle-class and upper-class flight to the suburbs has not been nearly so dramatic in European and Canadian cities, most of which remain attractive places in which to live and work.

World-wide trends toward greater car use and urban decentralization appear to result from underlying economic and technological factors. Increasing per capita incomes have made car ownership and suburban home ownership more affordable. Suburbanization of firms has been encouraged by changes in production and distribution methods, the growing dominance of service industries, the decline of traditional manufacturing, and increased locational flexibility of firms due to advances in communications and transportation (Altshuler *et al.*, 1979; Webster *et al.*, 1986; Banister and Berechman, 1993; Dasgupta, 1994).

Those economic and technological trends have been very similar in North America and Western Europe, however, and do not explain the vast differences in urban transportation and land use patterns just described. As discussed in the following sections, different government policies on transportation, housing and land use are responsible for much of the differences between the USA and Europe regarding car use and urban development.

URBAN TRANSPORT PROBLEMS

Every country suffers to some extent from transport problems, but the overall severity of the urban transport problem varies considerably, even within each individual country. In general, transport problems are most serious in the largest cities. In small and medium-sized cities problems certainly exist, but they rarely reach the crisis proportions found in large cities. That limitation of the transport crisis to the large cities is hardly much of a consolation, however, since large cities have been growing rapidly throughout the world, especially in developing countries. Large cities in Europe and North America, by comparison, have not been growing that fast, but their crucial economic, political and social functions make their transport problems extraordinarily important.

Congestion

Perhaps the most widespread transport problem afflicting the cities of Europe and North America is congestion. Fiscal austerity and environmental opposition have slowed down roadway construction in most countries, while increased car ownership, suburbanization and the diversion of rail freight to roads have raised the total demand for road use. The obvious result of the growing gap between demand and supply is traffic congestion. Of the 92 European cities and 14 North American cities surveyed in the OECD study, 70 per cent reported increasing congestion problems between 1970 and 1990, with average travel speeds having declined by about 10 percent (Sharman and Dasgupta, 1993). That is not surprising in light of the 60–70 per cent increase in car traffic volumes in the same cities during those two decades. Forecasts of another 30 per cent growth in car traffic by the year 2000 probably justify the expectation of more severe congestion in the coming years.

Environment

The problems of air pollution, noise, and other environmental impacts of urban transport are perhaps less immediate and less visible than the congestion problem, but they are probably even more important, especially over the long run. Congestion causes time delay. Environmental degradation, by comparison, leads to long-term health problems and destruction of the natural environment. Moreover, the environmental impacts of transport are acknowledged as international in scope, with the EU, for example, taking an official role in coordinating environmental policies among its member countries.

The objective extent of environmental problems from urban transport varies from country to country and from city to city. Moreover, the subjective perception of environmental problems varies perhaps even more. In Germany, Switzerland and Austria, for example, there is a virtual obsession with the environment, although the environmental problems there are not that much more serious than elsewhere in Western Europe, and are far less serious than in Eastern Europe. In Italy, France and England, by comparison, public concern about automotive air pollution and noise seems quite limited and has certainly not had a significant political impact, except for isolated responses to crisis situations, such as the ban on cars in the historical centres of a few Italian cities.

Energy

Energy use in transport appears to be a periodic problem, rising to the top of the policy agenda in years of sudden shortages and price increases and then largely ignored as supplies increase and prices fall. Thus, throughout Western Europe and North America, energy use was viewed as the most important transport problem during the energy crises of 1973 and 1979. Concern about energy use has diminished since then, both due to increasing energy supplies and technological advances in engine design that have improved the energy efficiency of automobiles and trucks.

Safety

Transport safety is a problem virtually everywhere, both in rural areas and in cities. Every year over 100 000 people are killed in traffic accidents in Europe and North America, and over two million are seriously injured. It is amazing how much traffic safety varies from one country to another. As shown in Table 2.7, the traffic fatality rate per 100 000 population ranges from 7.2 in Sweden to 44.8 in Hungary. Similarly, the traffic fatality rate per 100 million vehicle-km ranges from less than 1.5 in Finland, the Netherlands, the USA, Canada and the UK to 6.1 in Poland and 9.6 in Hungary.

Even among the most economically and technologically advanced countries of northwestern Europe and North America, traffic fatality rates vary widely, although in all those countries rates have fallen sharply from 1970 to 1992, largely due to better vehicle design and seat belt use. By contrast, fatality rates have risen alarmingly in the formerly Socialist countries of Eastern Europe as car use has burgeoned while roads, vehicle design and driver behaviour remain unsafe.

Finance

Transport financing has become an increasingly serious problem throughout Europe and North America, partly because of the rising costs of

Table 2.7 Traffic fatality rates in selected countries, 1970–92

	Per 100 000 population			*Per 100 million vehicle-km*		
Country	*1970*	*1980*	*1992*	*1970*	*1980*	*1992*
Austria	30.2	23.1	15.8	n.a.	4.0	2.8[**]
Belgium	30.4	24.3	18.7	9.2	5.2	3.3[***]
Canada	23.8	16.6[*]	13.8	4.0	3.8	1.4
Czechoslovakia	15.2	10.6[°°°]	14.4	n.a.	n.a.	n.a.
Denmark	24.4	13.5	11.1	5.0	2.7	1.5
Finland	22.5	11.5	11.9	6.0	2.0	1.4
France	32.1	25.2	15.9	8.3	4.6	2.0
Germany	31.6	21.2	11.6[***]	7.1	3.8	1.6[***]
Greece	11.8	14.1	17.4	n.a.	7.5[°°°]	n.a.
Hungary	16.5	15.2	44.8	12.0[°]	9.0[°°]	9.6[**]
Italy	20.3	16.2	13.2	6.1	3.3	1.9[***]
Japan	16.2	7.5	9.2	6.0	2.2	1.7
Netherlands	24.2	14.1	8.4	5.9	3.1[°°°]	1.3
Norway	14.4	8.4	7.5	5.0	n.a.	n.a.
Poland	10.6	13.4[*]	18.1	30.0	n.a.	6.1
Portugal	22.1	32.0	25.1[***]	17.3	9.8[°°°]	n.a.
Spain	16.1	17.3	15.4	11.3	7.0	4.8
Sweden	16.1	10.2	7.2	n.a.	n.a.	n.a.
Switzerland	27.0	20.0	12.7[***]	n.a.	n.a.	n.a.
UK	13.4	11.1	16.3[***]	3.6	2.1	1.1
USA	26.9	22.6	8.2[***]	3.0	2.1	1.2[***]

n.a. = not applicable.
[°]1971. [*]1984.
[°°]1978. [**]1990.
[°°°]1979. [***]1991.
Sources: Elaborated by the authors from Eurostat; IRF; DOT, *Transport Statistics Great Britain*.

building, maintaining and operating transport systems, and partly because of the overall fiscal austerity experienced by all government levels in virtually all countries. Detailed comparisons of fiscal problems among different countries are difficult because of misleading currency exchange rates, inconsistent accounting methods and differences in the extent and quality of transport services. The specific situation in individual countries will be examined in Chapters 3–10. In general, however, all the countries of Europe and North America face severe problems in funding transport projects, especially urban transport. That has led to a substantial decrease in real, inflation-adjusted subsidy funding during the decades of the 1980s and 1990s. Moreover, the lack of funds in the public sector has forced many countries to increase private sector involvement in the ownership, operation and financing of transport systems.

Equity

The equity problem in urban transport probably exists everywhere. The poor, the elderly and the handicapped suffer increasing mobility problems, especially as public transport services are curtailed and decentralizing land use patterns require ever more travel to reach employment, education, shopping and recreation destinations. The problem of unequal mobility is particularly acute in the USA, and it has led to considerable government action there to improve the mobility of disadvantaged groups, especially the elderly and the handicapped. In Canada and Europe, by comparison, the equity problem in urban transport has not evoked much public concern or policy action aimed specifically at improving mobility for certain groups. The transport problems of the poor, in particular, are viewed as deriving from the more basic problem of inadequate income, and public policies are thus more directed at income redistribution through general social assistance programmes than in the USA.

GOVERNMENT TRANSPORT POLICIES

Probably the most important difference between the urban transport policies of North America and Europe is the much higher cost of owning and operating an automobile in Europe. That higher price is due to high taxes, fees and user charges, partly to internalize the social and environmental costs of vehicle use, and partly as a convenient source of general government revenues. The cost of auto use in Canada is somewhat higher than in the USA, but still much lower than in Europe. The USA, however, distinguishes itself from Canada through its huge investment in urban and suburban highway systems, combined with the most extensive interurban highway network in the world. That roadway infrastructure was provided almost entirely by the public sector.

Road supply

The USA had 25 metres of road per capita in 1992 compared to a European low of 4 metres per capita in Spain and Greece and a high of

21 metres per capita in Norway (IRF, 1994). Even when the comparison is restricted to urban roads, the differences between the USA and Europe are large. Newman and Kenworthy (1989), for example, report an average of 6.6 metres of road per capita in the 10 largest American urban areas, about three times more than the 2.1 metres per capita in the 12 largest European urban areas, and 2.7 metres per capita in Toronto. In short, the USA has invested much more in its road network than Canada and most European countries.

Road finance

Perhaps even more importantly, the overall financing of road transport has much more heavily favoured automobile ownership and use in the USA than in Europe. Overall, road users in the USA pay only 60 per cent of the costs of road construction, maintenance, administration and law enforcement through taxes and user charges. The remaining 40 per cent (amounting to $30.7 billion in 1990) is subsidized through general government revenues. In contrast, road user taxes exceed government expenditures on roads in every European country. The ratio of road taxes to expenditures ranges from 5.1 in the Netherlands to 1.3 in Switzerland, but most European countries collect at least twice as much from road user taxes as they spend on roads (IRF, 1994). Thus, road users are heavily subsidized in the USA, whereas in Europe they pay such high road use taxes that they contribute significantly to overall government finance.

Total road user taxes per motor vehicle are about five times higher in Europe than in the USA, although there is much variation from one European country to another. Two important types of road user tax are the petrol tax and the sales tax on motor vehicles. The petrol tax per litre in Europe ranges from five to ten times the tax rate in the USA, resulting in prices that are two to four times as high (OECD, International Energy Agency, 1994). Most of the difference in petrol prices between Europe and the USA is due to differences in tax rates, not differences in the base price of petroleum. Differences in the sales tax rate on new cars are even more striking. Sales taxes on cars in the USA vary from one state to another, but they are all quite low, ranging from 5 per cent to 8 per cent. Most European countries have sales tax rates (including first registration fees) somewhere between 25 per cent and 40 per cent, with rates all the way up to 144 per cent in Portugal, 148 per cent in Finland and 180 per cent in Denmark (IRF, 1994).

Clearly, government taxation has made car ownership and use in Europe much more expensive than in the USA. That is partly the result of

EU sensitivity to the social and environmental costs of automobile use and the willingness of national governments to levy road user taxes in order to internalize those costs (ECMT, 1990). In addition, automobile taxation has long been a convenient source of general revenues for European national governments. Unlike the USA, successive increases in road user tax rates have been accepted without much political opposition. Perhaps that is because high tax rates were implemented in Europe before car ownership was widespread. The high cost of car ownership and use is generally accepted as a fact of life in Europe. In contrast, any attempt to increase petrol or motor vehicle taxes in the USA is viewed as an attack on the American way of life and is vigorously opposed by the vast majority of voters and by powerful special interest groups, such as the motor vehicle and oil industries and the American Automobile Association.

Parking policy

Of course, there are a number of other important government policies in the USA that have strongly encouraged automobile use and thus accelerated the decline of public transport. According to the *Nationwide Personal Transportation Survey*, drivers in the USA benefited from free parking for 99 per cent of all trips they made in 1990, and for 95 per cent of all work trips (Federal Highway Administration, 1992c).The widespread availability of plentiful free parking is largely the result of government policy. Free parking for employees has been treated for many decades as a tax-free fringe benefit for employees and as a tax-deductible expense for firms. Similarly, the cost of providing parking for customers is also tax-deductible. Those tax incentives for providing free parking have been reinforced by local government zoning laws and building codes that require a large supply of parking relative to square feet of office or retail space. The World Resources Institute estimates that free parking in the USA represents an implicit subsidy to car users of $85 billion per year (World Resource Institute, 1992).

The overall supply of parking is far greater in American cities than in Canadian and European cities. Newman and Kenworthy (1989) found that the 12 largest European cities had an average of 211 parking spaces per 1000 central city workers, roughly the same as the 198 spaces per 1000 workers in Toronto, but only half as many as the average of 380 parking spaces per 1000 central city workers in the 10 largest American cities. The large supply of mostly free parking in American cities obviously encourages car use.

Externalities

In addition, automobile drivers in the USA are implictly subsidized by the amount of the car's social and environmental costs. Two different studies by the World Resources Institute and the Worldwatch Institute both estimate that the indirect social and environmental costs of automobile use amount to at least $300 billion per year in the USA, or $2400 per automobile (World Resource Institute, 1992; Renner, 1988). The failure of public policy to internalize such external costs of car use unquestionably has made such use in the USA artificially cheap and thus has strongly encouraged ever more car ownership and use. The much higher taxation of ownership and use in Europe might be viewed as an internalization of at least some of the external costs of private-car use. Even if the main reason for higher taxes in Europe is to raise general government revenues, the impact of the higher taxes is the same, namely to raise the price of private-car use to more socially and environmentally optimal levels than in the USA.

Pedestrian transport

Not only do European governments tax car ownership and use more heavily than the USA, they also put more restrictions on car use. Pedestrian zones, for example, are far more widespread in Europe than in the USA. Many European cities have extensive, well-coordinated networks of car-free streets, providing a pleasant, safe environment for relaxation, shopping, socializing and walking for exercise. By contrast, the main car-free pedestrian environment in the USA is the suburban shopping mall, hardly a match for the interesting historic city centres of Europe. A few American cities have set aside one or two central streets as pedestrian malls, but their limited extent and the general unattractiveness of American central cities have hampered their success. Traffic calming is another European policy initiative aimed at improving pedestrian transport. Begun in the Netherlands, it has spread to much of Scandinavia, Germany, Austria and Switzerland. By reducing the volume and speed of car travel through residential neighbourhoods, traffic calming has greatly improved pedestrian and cyclist safety as well as reducing noise and air pollution.

Some European countries, however, have been unwilling to impose severe restrictions on vehicle use. In France, Italy, Spain, Portugal and the UK, car use in cities has been more accommodated than restricted (with occasional exceptions in specific cities, of course). Moreover, even where laws and regulations restricting car use have been passed in these European countries, they often remain unimplemented or unenforced. Parking bans and speed limits, for example, are so unpopular and so widely violated that many motorists do not take them seriously.

Technological solutions

The USA has relied primarily on technological fixes to the various problems of automobile use rather than trying to change travel behaviour through government actions. Since 1970, the USA has imposed increasingly strict regulations on automobile manufacturers to produce cleaner, quieter, safer and more fuel-efficient vehicles. Vehicle emissions standards, for example, were introduced in the USA almost two decades earlier than in Europe, and even today they remain much stricter than the standards in any European country. Whereas stringent federal standards have been introduced rather quickly in the USA without a great deal of bickering among the 50 states, the member countries of the EU seem to be taking forever to agree on even the mildest of uniform standards for vehicle design and fuels. Until 1986 each EC member country had a veto right, and achieving unanimity was often impossible. Even since 1986, small minorities in the EC Council of Ministers have been able to block any proposed changes in motor vehicle standards. Yet it is precisely in the area of vehicle design that almost all of the improvements in energy efficiency, safety and environmental impacts of automobile use have been made in the USA. In that respect, the USA is far ahead of Europe. Canada also lags behind the USA somewhat, but has been forced to adopt comparable standards due to the extreme interrelatedness of the car industries in the two countries. Similarly, environmental impact assessments for transport projects of all kinds are more advanced in the USA, and environmental standards are higher (Weiner, 1992). Indeed, some observers have argued that environmental regulations are too stringent in the USA, blocking or excessively delaying transport projects and increasing their costs (Altshuler *et al.*, 1979).

Both the USA and Europe are currently undertaking extensive research programmes aimed at computer optimization of motor vehicle travel on roads. The American version is called the Intelligent Vehicle Highway System (IVHS); the European versions are called 'Sirius', 'Carminat' established within the 'Prometheus' and 'Drive' research programmes of the EU. The initial stages of this new technology will provide drivers with information on routing possibilities and road conditions that could help both car and truck drivers save time and avoid accidents. More advanced stages may eventually allow automated vehicle operation. The 'smart car/smart road' technology is still in its experimental stages, however, and it is not certain whether it will really pay off.

Behavioral modification

Europe has been ahead of the USA in regulating travel behaviour as a means of improving urban transportation. Seat belt use laws, for example, were imposed in Europe 10–15 years earlier than in most American states. Likewise, licensing of drivers is much stricter in Europe, and the fees charged for driver training and licensing are much higher. Standards for drunken driving are stricter in most of Europe, enforcement is better, and the penalties for drunken driving are more severe than in the USA.

It is perhaps not surprising that the American approach has concentrated more on technological fixes to the urban transportation problem, whereas most European governments have been less hesitant to regulate travel behaviour directly, even if that has meant restrictions on individual freedom. Behavioral modification has been so unpopular in the USA that few politicians have had the courage to risk re-election by even proposing such measures. That probably explains the long delay in passing even the most obviously effective policies such as seat belt use laws. Of course, there is also a very strong pro-car lobby in Europe, but it has not been as successful as its American counterpart at preventing the adoption of policies restricting car use.

Public transport trends

Whereas public policies in the USA have generally made automobile use easy and inexpensive, public transport was allowed to deteriorate for decades. Only after the mid-1970s did public transport receive substantial financial support from federal, state and local governments, and by then, ridership had fallen to only a third of its 1945 level. In spite of truly massive subsidies (over $140 billion between 1975 and 1994), it has been possible to recapture only a minute portion of that lost ridership. Indeed, total public transport use grew less than 10 per cent over the entire period from 1975 to 1994. European countries, by comparison, generally did not allow their public transport systems to deteriorate after the Second World War. Indeed, many systems were completely rebuilt, expanded and modernized as war damage was repaired. Moreover, the slower growth in per capita incomes in Europe, together with the much higher cost of car ownership and use, ensured for public transport a larger pool of potential users than in the USA, where the car quickly became a deadly competitor.

Currently, however, public transport is experiencing problems everywhere in Europe and North America. Use has been falling in many countries (see Figure 2.2), and the financial crisis of public transport seems to become more desperate with each passing year. Tables 2.8 and 2.9 show that public transport is unprofitable throughout Europe and North America, but that the degree of unprofitability varies greatly, both among countries and among cities within the same country.

Nation-wide aggregate statistics on operating costs and revenues are only available for a few countries. As a whole, public transport covers less than a third of its operating expenses from passenger fares in Italy and the Netherlands, compared to 60 per cent in Germany (see Table 2.8). It is notable, however, that even in Germany, where public transport is least unprofitable, 40 per cent of expenses must be financed through subsidies. The range of variation is even more extreme among individual cities. For the 33 cities listed in Table 2.9, the ratio of passenger revenues to operating expenses ranges from only 17 per cent in Rome to 79 per cent in Berne (Switzerland) and 82 per cent in Porto (Portugal).

In Europe, Canada and the USA, the combination of overall fiscal crisis at all government levels and the enormous subsidy requirements of public transport have led to various attempts to increase productivity and reduce costs by encouraging more private sector involvement in public transport. As of 1994, however, public transport remained overwhelmingly in public

Table 2.8 Fare recovery ratio in urban public transport in selected countries (1992 or latest available year)

Country	*Ratio*
Canada	53
France	55
Germany	60
Italy	22–30
Netherlands	28
Sweden	44
USA	43

Note: Fare recovery ratio = passenger fare revenues as percentage of operating expenses.
Sources: Ministries of Transport or national public transport associations.

Table 2.9 Fare recovery ratio in selected urban areas (%)

Urban area/year	*Ratio*
Montreal (CA) 1988	40
Toronto (CA) 1988	68
Vancouver (CA) 1988	30
Atlanta (USA) 1990	35
Detroit (USA) 1990	24
Houston (USA) 1990	30
Philadelphia (USA) 1990	49
San Diego (USA) 1990	42
Barcelona (E) 1991	64
Porto (P) 1988	82
Bologna (I)* 1991	34
Genoa (I) 1988	25
Milan (I)* 1991	28
Rome (I) 1991	17–23
Vienna (A) 1991	50
Bruxelles (B) 1988	28
Liege (B) 1988	42
Amsterdam* (NL)	18
Oslo (N) 1993	60
Stockholm (S) 1988	30
Malmo (S) 1988	40
Helsinki (SF) 1988	40
Hamburg (D) 1988	60
Munich* (D) 1991	42
Copenhagen (DK) 1991	52
Bordeaux (F) 1993	38
Grenoble (F) 1993	62
Lille (F) 1993	67
Lyon (F) 1993	51
Paris (F) 1991	43
Berne (CH) 1991	79
Zurich (CH) 1991	66
Osaka (J) 1988	108

Note: Fare recovery ratio = passenger revenues as percentage of operating expenses.
*Municipal company only.
Sources: National public transport associations; RATP (1992); Quin *et al.* (1990).

As of 1994, however, public transport remained overwhelmingly in public ownership and control.

Public transport ownership and control

Table 2.10 provides summary information on the ownership of public transport systems in Western Europe and North America. Table 2.11 reports comparable information on the level of government control over public transport. Both tables are intended as rough guides. There are, in fact, ambiguous or borderline cases that are difficult to classify one way or the other. Moreover, it is rare that the situation is really all or nothing in any country; there are numerous exceptions to the generalizations in the tables. Finally, the situation can also be quite different according to specific mode of public transport. Rail services – both urban metros and suburban rail services – are almost always publicly owned, operated, and subsidized. Long-distance and suburban rail services in Germany were quasi-privatized in 1994 (with all corporate stock shares still owned by the federal government), and Britain and Italy have similar plans for their national railways. Full privatization has generally been limited to bus services, however, where infrastructure costs are much lower.

Privatization and deregulation

There are three important time trends that are not portrayed in Tables 2.10 and 2.11. First, private involvement has, in fact, been increasing in most countries during the 1980s and 1990s, although the current extent of private ownership, as a percentage of total public transport services, remains minimal. In the USA, the Reagan and Bush Administrations (from 1980 to 1992) enacted numerous programmes, stipulations and guidelines to encourage private ownership; yet by 1994, only 3 per cent of total public transport services were provided by privately owned systems. Only in Britain and the Scandinavian countries has privatization made important inroads. Most bus services outside London have already been completely deregulated and increasingly privatized. The partial deregulation of both urban and interurban bus services in Norway, Sweden, and Denmark has led to extensive competitive bidding and contracting-out of services to both private and public operators. By comparison, urban bus and rail systems in Germany, Austria, Switzerland, Belgium and the Netherlands remain completely public, subject to full regulation and government control over fare structure, routes and service standards. Moreover, the urban public transport systems in Germany, Austria and Switzerland are

Table 2.10 Status of ownership of urban public transport companies in selected countries

	Type of ownership		
Country	*Private*	*Public*	*Mixed*
Austria	X some rail	X	
Belgium		X	
Canada		X	
Denmark	X	X	
Finland	X		
France		X Paris	X
Germany		X	
Ireland		X largest cities	
Italy		X	X commuter rail
Netherlands		X	
Norway	X	X	
Portugal	X	X	
Spain	X	X largest cities	
Sweden	X Goteborg	X	
Switzerland	X some rail	X	
UK	X	X London	
USA		X	

Note: Suburban rail included.
Sources: Banister, Berechman and de Rus (1992); Tyson (1993).

mostly divisions of city-owned joint utility companies, with losses from their transport branches cross-subsidized by the profits of their other utility operations.

The main purpose of deregulation was to promote competition. The British chose to deregulate by increasing competition *on* the road, while most other countries chose to increase competition *for* the road. In the Netherlands, the system had been based on monopolies for many years. Each public transport operator had a specific contract with a city, but contracts were automatically renewed. Starting in 1995, however, the central government will put some public transport services out to competitive tendering. All European transport companies are invited to compete for the right to operate. Once a company has won the contract, it has the monopoly for the service. A similar system is in operation in Sweden, Norway and Denmark. County public transport authorities must invite public and private operators to bid for contracts. Those who win have monopoly rights to provide services on specific lines. In Scandinavia, therefore, operators have line-by-line contracts while, in the Netherlands,

Table 2.11 Who is responsible for urban public transport in selected countries?

	Government level		
	National	*Intermediate*	*Local*
Austria		Länder	municipalities
Belgium		regions	
Canada		provinces	municipalities
Denmark		counties	2 areas only
France			groups of municipalities
Germany			municipalities
Italy		regions	municipalities or groups of communes
Netherlands	central government		
Norway		counties	
Spain		regions*	municipalities
Sweden		counties	
Switzerland		cantons	municipalities
USA			municipalities, authorities and special districts

**Communidades autonomas* (regional level governments).
Notes: local means municipal, intermediate means county, province or region. Suburban rail is excluded from this table.
Although in many countries and urban areas, public transport policy is decided and implemented by joint authorities, we have tried to state which institution is primarily responsible for public transport policy (the organizing authority). By policy, we mean planning, management, use of financing, setting fares and awarding franchises. Of course, these various functions are often spread out among various governmental tiers. We refer here at the tier which is the most involved in the policy-making and implementation. The reader will find more detailed analysis in Chapters 3–10.

contracts are on a system-wide basis. Deregulation in Europe has generally not gone beyond that, except in Great Britain, of course, where competition has been enhanced by free-market entry and exit, elimination of most government control over fare structure and service levels, and extensive contracting-out via competitive bidding for those services not subject to on-the-road competition.

Similarly, much privatization appears to be occuring more in word than in action. In many cases, it simply means the transfer of public companies from public to private corporate legal status, as opposed to fully private ownership and control. Of course, even the more limited transformation may be significant. It forces companies to balance their budgets. Perhaps more importantly, their employees are no longer civil servants; thus their wages, fringe benefits and work conditions generally become less generous, and their employment is less secure. In short, even limited privatization should increase pressure for cost control and productivity gains.

Devolution of policymaking

A second important time trend not revealed by the two tables is the devolution of control, management, planning and finance of public transport from central (national) government levels to state (provincial) and local (municipal) government levels. With the exception of the Netherlands, the central governments have little control over the organization, route networks and service standards of urban public transport. Central governments still contribute substantially to public transport finance, but their subsidies account for an ever smaller percentage of total funding, as state and local governments are forced to bear increasing portions of the subsidy burden. That devolution of financial responsibility has been one of the main incentives for experimenting with deregulation and privatization. With declining central government subsidies, state and local governments have become increasingly sensitive to the subsidy burden of public transport.

Regional coordination

The third time trend is the spreading regionalization of public transport, with increasingly integrated services and financing over entire metropolitan regions and conurbations. During the 1970s and 1980s, virtually all American urban areas adopted metropolitan-wide regional transit districts, encompassing many municipalities, counties and even states within the same district. During roughly the same period, German public transport systems were also increasingly integrated and coordinated through various forms of associations, with different degrees of cooperation in service, fare structure, subsidy, planning and management. In France, local public transport has been increasingly organized through groups of municipalities, sometimes associated with their *départements* (counties). Not all countries are following this trend towards regionalization, however. In Italy, for example, local governments have been unable to get together to form metropolitan public transport authorities. Moreover, metropolitan transport authorities (Passenger Transport Authorities, or PTAs) were abolished in Great Britain along with metropolitan counties.

All these developments suggest rather uncertain times for public transport in the years to come. After the expansionism of the 1970s and the consolidations of the 1980s, the decade of the 1990s so far has brought considerable austerity, and the likelihood is that financial pressures will worsen in the coming years. Increased productivity through selective privatization and deregulation may help somewhat, but in the current environment of rapidly rising car ownership and use and continuing suburbanization, public transport will be in a very difficult predicament indeed.

Privatization of motorways

Just as fiscal austerity has slowed down investments to public transport systems, it has also curtailed expansion of highway networks. And just as subsidy cutbacks have provoked deregulation and privatization measures in public transport, they have also motivated a search for private funding for highways. In the USA, for example, numerous private motorways are currently being built or planned, with the costs financed by the anticipated toll revenues from users. Similarly, most of the planned extensive motorway construction in Poland, Hungary and the Czech Republic is dependent on private finance. Although the objective need for more highway capacity is unquestionably greater in Eastern Europe than in Western Europe or North America, public funding for the necessary investment is virtually non-existent. Unless private investors are willing to fill the funding gap, highway expansion in Eastern Europe may be delayed for decades.

Travel demand management

Both fiscal austerity and environmental opposition are blocking major expansions to highway capacity. Yet travel demand continues to grow due to increased car ownership and suburbanization throughout Europe and North America. Given the limitations of supply-side approaches to solving the urban transport problem, policies are increasingly turning to demand-side solutions. Instead of catering to whatever travel demands arise, governments are starting to adopt policies aimed at limiting or rechannelling travel demand. For the most part, this has involved measures to encourage increased car-pooling or public transport use, but it can also entail redistribution of car travel from peak periods to off-peak periods. In the USA, for example, over 20 cities have established express lanes for high occupancy vehicles (HOVs) on about 60 key arterial highways. A few cities have also implemented selective ramp metering on freeways to give priority access to HOVs. In Europe, demand control measures have focused more on banning private cars from central city districts, limitations on parking, increases in parking fees, traffic calming in residential neighbourhoods, and the overall policy of high taxes on ownership and use.

Congestion pricing, another demand side policy, continues to be a popular notion among some transport analysts, especially economists, but in the three decades since it was first proposed it has made almost no headway. Having to pay for using scarce road capacity at peak hours may make a great deal of sense to the experts; charging a zero price obviously encourages overuse. Nevertheless, nowhere in Europe and North America have politicians been willing to endorse its implementation on a widespread basis. Even limited experiments have been rare. Congestion pricing is generally viewed by politicians, consumer groups, and the public at large as yet another form of taxation, an infringement on personal freedom and an unfair rationing of mobility.

A more popular pricing strategy has been the widespread use of deeply discounted monthly tickets for public transport, increasingly marketed as environmental tickets, combining the advantages of cheap travel and a clear conscience. Especially in Germany, Switzerland, Austria and the Netherlands, such environmental tickets have succeeded in increasing passenger volumes and, in some cases, even reducing car use, albeit at the cost of more subsidy. In coordination with higher taxes and fees for car use, reductions in public transport fares are intended to shift modal choice away from single-occupant vehicle use. The financial crisis in the public sector is limiting the use of 'carrot' approaches, such as public transport fare discounts, and forcing a shift to 'stick' approaches, such as higher taxes and fees. Thus, petrol taxes were raised drastically in Switzerland and Germany in 1993 and 1994 (by roughly US$0.50 per gallon), and parking fees have been rising rapidly throughout Europe.

Land use policies

We conclude this brief overview with a few comments on land use policies, which have been crucial to shaping urban development patterns and thus travel demand. Land use policies vary greatly from one country to another. By tradition, the Netherlands, Germany, Switzerland and the Scandinavian countries have very restrictive laws and regulations on land use. In contrast, Italy, Spain and Portugal have no significant government control over the use of privately owned land. France and Britain lie between those two extremes, with many land use plans and regulations, but few legal sanctions to enforce them seriously. Within North America, land use policies also vary. Historically, the USA has generally allowed private landowners to do whatever they want with their land, and efforts to control the use of private land have been opposed as infringements on basic freedoms and individual rights. Thus, urban and especially suburban developments have arisen rather haphazardly, as private developers, builders and land speculators seek to maximize their profits with little regard, if any, for long-run social and economic consequences. Land use controls tend to be much stricter in Canada, and there is much more government intervention (through subsidies, zoning, building codes and tax policies) to encourage more compact urban development. The strict land use controls in some countries obviously encourage more clustered, higher density development, while the lax controls in other countries invite suburban sprawl. Since public transport, and to some extent walking and bicycling, require compact development in order to be feasible, the variation among countries in land use controls has important implications for the evolution of urban travel behaviour.

So far, we have described the range of government policies in Europe and North America. Of course, it is equally important to evaluate those policies and to determine why certain policies succeeded or failed. In the course of our descriptions, we have already indicated in some cases the outcomes of policies. Chapter 11 contains a more systematic evaluation of policies, based on an in-depth analysis of urban transport in each of the eight chapters which follow.

CONCLUSIONS

Car travel and suburbanization are increasing in virtually all countries of Europe and North America, but current levels of use and suburbanization remain very different from one country to another. In particular, levels of car ownership and use are much higher in the USA than in Canada and Europe, and the low density suburban sprawl surrounding American cities remains one of their distinguishing characteristics.

Government policies have greatly affected developments in urban land use and transportation, and largely account for the much more car-oriented travel behaviour and lifestyle of Americans. The cost of ownership and use in Europe is far higher than in the USA, and the difference is due primarily to higher taxes, fees and user charges imposed on private car drivers. Likewise, European cities generally offer far superior transport alternatives to the car: not only more extensive public transport but also better possibilities for cycling and walking. Finally, land use and housing policies in Europe have either discouraged urban sprawl directly or at least provided fewer incentives for low density residential and commercial developments at the urban fringe.

Even the much lower levels of car use in Europe have caused severe problems of congestion, pollution and safety. The social and environmental problems of car use are all the more serious in Europe because of higher population densities, less available land and cities that are far less suited or even adaptable to automobile use. The problems of such use have arisen later in Europe than in the USA because levels of ownership and use in Europe have been about two decades behind developments in the USA. Now that car use is rising rapidly in Europe, its adverse social and environmental impacts are becoming ever more apparent.

There appears to be a consensus among voters and politicians in most European countries on the need to control automobile use to account for its negative impacts. That probably ensures the continuation of high taxation of car ownership and use and strict land use controls. It thus seems unlikely that Europe will ever see the extremely one-sided domination of transportation by the automobile or the endless low density suburban sprawl common in the USA. Nevertheless, it is virtually certain that car ownership and use in Europe will continue to grow, and that European cities will continue to decentralize. Worsening transport problems will be the inevitable result unless stringent policies are adopted to control car use. Immediate action is necessary to avoid an even more serious urban transport crisis in the years to come.

We now turn to our detailed analysis of urban transport in each of eight chapters: Germany, France, the Netherlands, Italy, Great Britain, Eastern Europe, Canada and the USA.

3 Germany: The Conflict between Automobility and Environmental Protection

Germany is a country full of contradictions, and that is also true in passenger transport. Prior to reunification in 1990, West Germany had the highest rate of car ownership in Europe, and the second highest in the world, surpassed only by the USA. East Germany had by far the highest rate of car ownership of any Socialist country. Since reunification, Germany continues to have more cars per capita than all but a few countries. Moreover, its system of limited access superhighways, the autobahns, is the most extensive in Europe and is the only highway system in the world without a general speed limit. Indeed, one sometimes has the impression that the autobahns are racetracks rather than highways. Repeated attempts to legislate a general speed limit for the autobahns have been vigorously opposed by various interest groups and the majority of German voters.

The Germans themselves – both in the East and the West – readily acknowledge their love affair with the automobile and their virtual obsession with car travel. By permitting extraordinary levels of mobility, flexibility, privacy and comfort, the car has enjoyed great popularity as a mode of transport throughout the world. Perhaps equally important in Germany, however, is the crucial symbolic value of the car. For most Germans, the car has long represented political freedom, general economic prosperity and personal socioeconomic status. That additional psychological attraction has further enhanced the car's lure. Germans even have a saying to express their association of the car with political and economic freedom: 'Freie Fahrt für freie Bürger!' (loosely translated: 'Unlimited car travel for free citizens!'). That battle cry is invoked by German car drivers every time an attempt is made to impose a speed limit on the autobahns. In short, Germans appear to be at least as fanatical as Americans in their addiction to the automobile. Indeed, whereas most Americans have no alternative to the car and thus use cars as a practical necessity, Germans freely choose cars even though they have a range of attractive and much less expensive alternatives.

As a counterpoint to that automania, Germany has had the strongest and most organized political opposition to the car in Europe, perhaps even in the world. The Greens have become a formidable political force, advocating environmental protection and conservation at every government level. Having begun as a grassroots movement around 1980, the Greens developed into a formal political party that has unceasingly tried to block highway construction, limit car use and promote public transport, bicycling and walking as the only environmentally responsible alternatives to automobile travel. Environmental concern, however, is by no means limited to members of the Green Party. The devastating consequences of excessive car travel in such a densely populated country have led to widespread environmental advocacy among the German population. The Social Democrats and even the Christian Democrats have enacted a range of measures at various government levels to mitigate the harmful impacts of car use.

Germany presents an interesting juxtaposition of Europe's oldest and strongest anti-car environmental movement with Europe's highest rate of car ownership and use. The love-hate relationship with the car is probably more passionate in Germany than anywhere else in the world. The conflict between automobility and environmental protection permeates virtually every aspect of German transport policy.

This chapter deals mainly with urban transport in West Germany. Transport policy in the unified Germany is at any rate very similar to policy in the former West Germany. Moreover, urban transport systems and travel behaviour in the former East Germany are rapidly approaching West German norms. The history of urban transport in East Germany during four decades of Communism is discussed in detail in Chapter 8 on the formerly Socialist countries of Central and Eastern Europe. Likewise, changes in the eastern portion of Germany since reunification in 1990 are only briefly summarized here, with a more detailed discussion in Chapter 8. Nevertheless, the present chapter acknowledges differences between the eastern and western parts of Germany in their transport systems, travel behaviour and investment needs. The main purpose is to portray the urban transport situation in the unified Germany, including the continuation of previous West German trends (such as privatization and devolution) and the modification of policies to deal with the new transport needs of a larger Germany.

TRENDS IN TRAVEL BEHAVIOUR

Car ownership and use have increased dramatically in Germany over the past four decades (see Table 3.1). In 1950, West Germany had one of the Western world's lowest rates of ownership: only 12 cars per 1000 population. Between 1950 and 1992, however, ownership skyrocketed to 492 cars per 1000 population, a 41-fold increase, giving Western Germany the second highest rate of car ownership in the world (Heidemann *et al.*, 1993; German Ministry of Transport, 1993a).

As shown in Table 3.1, the motorization rate of the unified Germany in 1992 was somewhat lower than the rate for West Germany alone (470 as against 492 cars per 1000 population). Although car ownership in the eastern part of Germany has almost doubled since the fall of Communism (from 237 to 447 cars per 1000 population), it is still about 10 per cent lower than the West German rate, thus bringing down the average for Germany as a whole. Nevertheless, even the unified Germany has one of the highest rates of car ownership in the world.

As car ownership has increased, so have vehicle kilometres of use. The data series for vehicle kilometres in West Germany starts in 1952. In the 40 years from then until 1992, car use rose from 18.2 billion km to 409.8 billion km, a 23-fold increase. Of course, the extremely rapid growth in car ownership and use far exceeded population growth, which was only 36 per cent for the entire 42-year period from 1950 to 1992 (excluding the increase due to the annexation of East Germany).

Corresponding to the trends towards ever-greater car ownership and use, the number of passenger km travelled by car also increased greatly between 1950 and 1992, and the percentage of motorized travel by car has continually risen. As shown in Table 3.2, passenger km by car in West Germany rose from only 31 billion in 1950 to 613 billion in 1992, roughly

Table 3.1 Trends in German population, car ownership and use, 1950–1992

	*1950**	*1960**	*1970**	*1980**	*1990**	*1992**	*1992*†
Population‡	47 696	55 958	60 651	61 566	63 726	65 100	80 800
Cars‡	570	4 490	13 941	23 192	30 685	32 007	37 947
Cars per 1000 population	12	80	230	377	482	492	470
Car km travelled§	18.2¶	73.2	201.1	297.4	401.6	409.8	470.9

*Includes only Western Germany
†Both Western and Eastern Germany.
‡Thousands of units.
§Billions of units.
¶For 1952.
Source: German Ministry of Transport, *Verkehr in Zahlen* (1982–1993).

Table 3.2 Trends in car travel and public transport use, 1950–92

	1950*	1960*	1970*	1980*	1990*	1992*	1992†
Passenger km by car‡	31.1	155.2	352.3	472.5	596.3	612.6	717.2
(% of total motorized travel)	35.5	64.9	78.5	80.4	84.4	84.1	83.6
Passenger km by public transport‡	56.5	83.8	96.5	115.1	109.8	116.2	140.5
(% of total motorized travel)	64.5	35.1	21.5	19.6	15.6	15.9	16.4
Urban public transport trips§	5 144	6 603	7 015	7 652	6 873	7 296	9 148
Urban public transport trips per capita	108	118	116	124	108	112	113

Notes:
*includes only Western Germany.
†both Western and Eastern Germany.
‡billions of units. §Millions of trips.
Source: German Ministry of Transport, *Verkehr in Zahlen* (1982–93).

a 20-fold increase in total car travel, and a 14-fold increase in vehicle kilometres of travel per capita.

During the same period, public transport use in West Germany was also growing, but much slower than car use (see Table 3.2). Total passenger kilometres by public transport only doubled (104 per cent increase), and passenger kilometres per capita increased by only 49 per cent, compared to the 1348 per cent increase in car travel per capita. As a consequence, public transport's proportion of total motorized ground transport in West Germany fell from 65 per cent to 16 per cent, while the car's share rose from 36 per cent to 84 per cent.

Those figures include intra-urban, interurban and rural travel. Unfortunately, corresponding statistics on vehicle kilometres and passenger kilometres of urban travel alone are not available, but similar statistics for other countries suggest that trends in total travel are indicative of urban travel trends.

For example, the total number of urban public transport trips in West Germany increased modestly between 1950 and 1980 (+49 per cent), then decreased from 1980 to 1990 (–10 per cent), and rebounded slightly from

1990 to 1992 (+6 per cent). Much of the change in public transport usage was evidently due to overall population trends, since on a per capita basis public transport usage in 1992 was only slightly different from 1950 (112 as against 108).

Table 3.3 shows modal split trends for all urban travel in West Germany. The figures were derived from a representative sample survey conducted for the German Ministry of Transport.

As a percentage of travel by all modes (including non-motorized travel), the car's share rose from 45 per cent in 1976 to 53 per cent in 1989. Public transport's share fell slightly from 13 per cent to 11 per cent. As in most European countries – and in sharp contrast to North America – bicycling and walking are important urban travel modes. Nevertheless, their relative significance has also fallen as urban areas in West Germany have begun to spread out and trip distances have increased. The combined modal split share of cycling and walking declined from 42 per cent in 1976 to 36 per cent in 1989.

As described in detail in Chapter 8, the direction of trends in Eastern Germany has been similar but with different timing. The car became relatively more important in urban transport over the entire period from 1972 to 1992 for which detailed surveys are available (see Table 3.4). The rise in car modal split was gradual, however, until the end of Communist dictatorship in 1989. It increased from 17 per cent to 25 per cent in the 15-year period from 1972 to 1987 and then from 25 per cent to 34 per cent in the four years from 1987 to 1991.

Public transport, by comparison, fell from 27 per cent to 19 per cent of total urban travel, and cycling and walking combined fell from 57 per cent to 47 per cent. Thus, the eastern portion of the now unified Germany remains less car-oriented than the western portion, but it is rapidly moving

Table 3.3 Modal split trends for urban travel in Western Germany, 1976–89 (as percentage of trips)

Year	*Car*	*Public transport*	*Bicycle*	*Walking*
1976	45.0	12.8	8.6	33.6
1982	47.0	13.0	10.2	29.8
1986	49.8	11.8	10.4	28.0
1989	52.5	11.1	9.8	26.6

Source: German Ministry of Transport, *Verkehr in Zahlen* (1982–93).

Table 3.4 Modal split trends for urban travel in Eastern Germany, 1972–91 (as percentage of trips)

Year	*Car*	*Public transport*	*Bicycle*	*Walking*
1972	16.6	26.8	10.6	46.0
1977	18.8	28.7	8.8	43.8
1982	20.1	26.8	9.7	43.3
1987	24.6	26.6	9.6	39.2
1991	34.3	18.8	9.4	37.5

Source: Förschner and Schöppe (1992).

towards more car use and less public transport. Indeed, in the first four years since German reunification, car ownership in Eastern Germany almost doubled, and public transport ridership fell by 56 per cent. The modal shift in Eastern Germany has been so rapid that urban travel differences between Eastern and Western Germany will probably become insignificant over the next decade or so.

URBAN LAND USE PATTERNS

With 222 people per square kilometre, Germany is one of the most densely populated countries of Europe, surpassed only by Belgium and the Netherlands. In addition, Germany is a highly urbanized country, and its cities and urban agglomerations are quite compact and densely populated. The largest German cities, however, do not appear to be any more densely populated than other large European cities. In their comparative study of 32 world cities, Newman and Kenworthy (1989) found that population and employment densities for central business districts, inner residential areas and suburbs were about the same for German and other European cities. The most important difference is that much more of Germany's land area is taken up by urban uses, leaving less land for agriculture, forests, and wilderness areas. That scarcity of land has led to a strict overall land use policy for the country, which also carries over to urban land use planning.

For example, the state *(Länder)* governments, in particular, sharply restrict new residential development at the urban fringe. Much privately owned land around cities is zoned exclusively for agriculture, forests, nature preserve or simply open space. Those restrictions greatly limit the supply of land available for urban development, driving up the price of land and thus encouraging quite dense development. The compactness of urban development, even in the suburbs, obviously facilitates public transport, which relies on high-volume, focused travel corridors.

Nevertheless, the trend towards decentralization found throughout Europe and North America can also be found in Germany (Heidemann *et al.*, 1993; Jansen, 1992). Increasingly, one finds shopping centres and office complexes near the edges of cities rather than in the centre, and German suburbs are generally growing faster than their central city counterparts. The 1993 OECD study of urban travel and land use trends from 1970 to 1990 included six German cities: Dusseldorf, Freiburg, Heidelberg, Berlin, Lübeck and Schwerin. All the German cities reported an increasing proportion of their metropolitan area population living in the suburbs, and a declining proportion of metropolitan area jobs located in the central business district (Sharman and Dasgupta, 1993).

The increased decentralization of German cities makes the provision of public transport more difficult. As German cities begin to spread out into the surrounding areas, trip lengths increase and radial traffic focused on the city centre becomes relatively less important. Cross-commuting is increasing, especially within some of the large urban agglomerations such the Rhine-Ruhr region, Frankfurt, Munich and Hamburg (Heidemann *et al.*, 1993; Jansen, 1993). The relative advantage of car travel is greatest precisely for those sorts of trips from one non-central location to another. That is probably one explanation for the rapid growth in car use and the stagnation in public transport use.

Eastern German cities have a rather different structure. Suburbanization under Socialism was driven not by the market but by the decisions of ministry bureaucrats and planners who located almost all new residential construction at the outermost periphery of the city, where land was most available. Thus, in the former East Germany – just as in Poland, Hungary, the former Czechoslovakia and the former Soviet Union – massive, high density apartment complexes ring virtually every city. Because there was almost no coordination between housing and industrial location, the journey to work was quite long and time-consuming. Moreover, the peripheral apartment complexes were never adequately supplied with recreational, educational, medical and shopping facilities, so that long trips into the city centre were necessary for those purposes as well. The current trend in Eastern Germany is towards ever more commercial facilities on the fringe to service the population that lives there. That will probably reduce trips to the central city and greatly increase the amount of cross-commuting between suburbs. Both developments are likely to encourage more car use and less reliance on public transport.

URBAN TRANSPORT PROBLEMS

Urban transport problems in the eastern and western parts of Germany were dramatically different prior to reunification in 1990, but have become quite similar since then. The current problems can be classified into two categories: (1) the harmful social and environmental impacts of car use and (2) the huge subsidy requirements of public transport.

Social and environmental costs of car use

As elsewhere in the industrialized world, the automobile and truck have caused enormous social and environmental problems: traffic accidents, air pollution, noise, destruction of farmland and open space, disruption of urban neighbourhoods, energy waste and congested roads.

Western Germany has a longer history of such problems simply because it has a much longer history of high levels of car use. Before reunification, car ownership rates in the East lagged behind those in the West by about two decades. Levels of car travel per capita in the East lagged even further behind because cars were used infrequently and mainly for social and recreational travel on weekends and holidays rather than for the daily journey to work.

Of course, even with their much lower levels of car use, East German cities had to deal with increasingly serious transport problems. The primitive East German cars (Trabants and Wartburgs) produced much more pollution per kilometre driven than West German cars. In combination with the extremely severe industrial pollution in East German cities, car emissions were becoming an increasingly serious problem as ownership and use grew even at modest rates under Socialism. The meteoric rise in car use in Eastern Germany since reunification has exacerbated the environmental problems there. Similarly, East German roads were very unsafe, and East German cars were dangerous to drive, but the cars were so slow, and travel so limited, that East German traffic fatality rates per capita were much lower than those in the West. Since reunification, however, traffic fatality rates in the East have more than doubled, and now considerably exceed those in the West (Pucher, 1994).

As shown in Table 3.5, car travel in Western Germany has become progressively safer since 1960. Traffic deaths per 100 million passenger

Table 3.5 Traffic accidents in Germany, 1960–92

Year	*Injuries*	*Deaths*	*Deaths per 100 million passenger-km*	*Deaths per 100 000 population*
1960	454 900	14 406	9.3	25.7
1970	531 800	19 193	5.4	31.6
1980	500 500	13 041	2.8	21.2
1990	448 200	7 906	1.3	12.4
1992	425 500	7 294	1.2	11.2
1992*	516 600	10 627	1.5	13.2

*Eastern and Western Germany.
Source: German Ministry of Transport, *Verkehr in Zahlen* (1982–93).

kilometres fell from 9.3 in 1960 to 1.2 in 1992. In spite of the large increase in car travel, even the death rate per capita has fallen, from 26 per 100 000 population in 1960 to 11 in 1992.

Nevertheless, the 7294 traffic deaths and 425 500 injuries in Western Germany in 1992 constitute a serious safety problem. Traffic safety has become an even more serious problem in Eastern Germany, which now has a per capita fatality rate more than double the Western German rate (23.4 as against 11.2 deaths per 100 000 population).

Two aspects of the car's environmental impact have raised particular concern in Western Germany: air pollution and the encroachment of roads on agricultural land, forests and undeveloped land in general. The automobile is the main source of carbon monoxide (CO), nitrogen oxide (NOx), and hydrocarbon (HC) pollution and has also contributed to airborne lead and particulate pollution. Germans are especially sensitive to the key role of car emissions in causing the acid rain responsible for the high percentage of trees in Germany's forests which are dead, terminally ill or seriously damaged. A high population density, the scarcity of open spaces and an unmatched passion for hiking have made the dying forests (*Waldsterben*) an extremely important political issue (Kunert, 1988; Pucher and Clorer, 1992).

Technical improvements in engine design have succeeded in reducing several types of pollution per kilometre driven, but the increase in the amount of automobile travel has largely offset those per kilometre reductions. In Western Germany, for example, total NOx emissions increased from 0.8 million tons in 1970 to 1.4 million tons in 1980 and 1.7 million tons in 1990. Similarly, HC emissions increased from

0.9 million tons in 1970 to 1.2 million tons in 1980 and 1990. Total CO emissions rose slightly from 1970 to 1980 (8.4 to 8.5 million tons), then fell significantly by 1990 (6.1 million tons: German Ministry of Transport, 1993a).

Environmentalists have opposed almost all expansions to the existing road network. In rural areas, new highways claim agricultural land, wetlands and forests, and generally disrupt both the natural ecological system and the rural, agricultural lifestyle of the population. In urban areas, new or expanded roads often disrupt communities, induce excessive traffic (and the noise, pollution and congestion that causes) and encourage sprawled suburban development.

Whether in cities, suburbs or rural areas, there is a constant battle in Germany over road construction policy. On the one side, automobile enthusiasts and industry representatives want to expand the road network to maximize the benefits of car ownership. On the other side, environmentalists and community activitists want to preserve the natural environment and their residential communities at all costs. The battle over roads has spread rapidly to Eastern Germany, where several key links in the highway network are missing, but can only be completed at the cost of significant environmental damage and intrusion into the rural landscape. In urban areas, the fear is that more roads will further accelerate the already massive modal shift away from public transport towards the environmentally harmful car.

Of course, there are many other social and environmental problems of car use. As elsewhere in Western Europe and North America, the energy shortages of 1973 and 1979 induced sudden but short-lived concern with energy use in transport. The fuel efficiency of new cars barely increased at all in Germany between 1970 and 1992 (from 10.2 to 9.8 litres of petrol per 100 km driven). Fuel savings from engine improvements were offset by the shift to larger, more powerful cars. Energy use is definitely not a high priority issue in Germany, although the country must import almost all its petroleum and thus is dependent on foreign suppliers.

Road congestion is widely viewed as a problem, but mainly by car users. Moreover, congestion is restricted mainly to key transport nodes at peak hours of travel and is not the pervasive problem one observes in many American cities. Congestion is generally more severe on the autobahns between cities than on urban roads. Indeed, traffic jams on the autobahns often extend over distances up to 100 km during peak holiday travel.

Finally, the equity problem in urban transport is not very serious in Germany and has low political priority. Unlike the USA, a wide variety of transport alternatives are available to those who cannot afford a car or who are physically unable to drive. Moreover, the federal and state governments provide deeply discounted ticket prices for students, the elderly and the handicapped. Of course, there are some destinations that can be reached much more conveniently by car, especially in the suburbs. Nevertheless, through some combination of public transport, cycling and walking, Germans can reach virtually any part of their cities. Both the compact land use pattern and the extensive transport alternatives to the car make German cities highly accessible to the entire population.

Financial problems

Financing public transport has long been an enormous burden on all government levels in West Germany. In Communist East Germany, the burden was borne exclusively by the central government, but, since reunification in 1990, the new Eastern German states (Länder) have been forced to adopt the West's multi-level system of subsidy finance.

In West Germany, the total capital and operating subsidy to all forms of public transport was DM 4.0 billion in 1970 and grew to DM 11.8 billion by 1980, and to DM 14.5 billion by 1990 (equivalent to about US$9 billion). Even adjusted for inflation, that represents a 65 per cent real increase in the total subsidy required for urban public transport (Pucher, 1994).

Unlike the highly centralized subsidy programme for all forms of public transport in the former East Germany, the West German approach has long been more decentralized and more differentiated by type of subsidy and mode of public transport. For example, there is a large difference between the financing of bus, tram and metro services and the financing of suburban rail services. The national railway in Germany (Deutsche Bundesbahn, or DB) provides both long-distance services and short-distance, suburban rail services. The federal government, which owns the railway, bears full financial responsibility for covering all operating deficits and investment costs. In 1990, it paid DM 4554 million (about $2.8 billion) just to offset the operating deficit of suburban rail services. That represented 65 per cent of the federal government's total financial contribution to public transport. Of the remaining DM 2444 million, 20 per cent went to a special capital subsidy fund for investment in all forms of urban public transport (including buses and metros), 9 per cent was for reduced fare subsidies for school pupils and the handicapped, and 6 per cent financed the reduced sales tax rate charged to public transport firms.

The total public transport subsidy coming from the state (*Länder*) level of government in West Germany was DM 3074 million in 1990. Of that total, 74 per cent was for financing reduced fares for school pupils and the handicapped. The remaining 26 per cent was for capital subsidies. The local level of government contributed DM 4443 million. Roughly 80 per cent of the local subsidy went to cover the operating deficit of bus, tram and metro systems, which are owned by municipalities as part of their overall utility operations. The local contribution is mainly in the form of a cross-subsidy from the profitable gas, electric and water utilities. About 10 per cent of the local subsidy is for capital investment, and another 10 per cent is for reduced fares to special groups (German Ministry of Transport, 1992). Overall, then, the West German federal government has been responsible for suburban rail financing and has contributed significantly to capital subsidies for other forms of public transport as well. State governments have focused on financing reduced fare programmes and capital investment, while local governments have borne virtually all the operating subsidy burden of bus, tram and metro services.

In Socialist East Germany, by comparison, the central government had full responsibility for financing all public transport services throughout the country. Passenger fares covered less than 20 per cent of the total operating costs of bus, tram and metro services in 1989, in contrast to over 60 per cent cost coverage through fares in West Germany. Suburban railway services operated by the East German national railway, the Deutsche Reichsbahn, were even more unprofitable, covering only 15 per cent of operating costs through fares, compared to 30 per cent cost coverage for suburban rail in West Germany. Thus, even if subsidies to public transport services in Eastern Germany are cut back to the fairly generous West German levels, substantial subsidy reductions would be required.

Currently, local public transport systems in Eastern Germany are barely managing to remain solvent, in spite of drastic fare increases and emergency aid from state and local governments. The total operating deficit of public transport in Eastern Germany (including commuter rail) amounted to DM 4.4 billion (about $2.8 billion) in 1992, for an average of DM 2.45 ($1.53) per passenger trip. Capital subsidies for 1992 are estimated to be roughly DM 3 billion (about $2 billion: German Ministry of Transport, 1992c). Since German reunification, the German federal government in Bonn has assumed full financial responsibility for the suburban rail services of the Deutsche Reichsbahn in Eastern Germany, just as it does for suburban rail services of the DB in West Germany. Moreover, the federal government has been providing temporarily high matching rates and extra funding levels to the new Eastern German states for capital subsidies to finance investment in bus, tram and metro systems. Compared to a maximum federal subsidy share of 60 per cent for investment projects in Western Germany, the federal government provided the full capital subsidy amount in Eastern Germany in 1991 (without any state and local match). The matching share was reduced to 90 per cent in 1992, and is eventually scheduled to be reduced to the normal Western German level.

In sharp contrast to the socialist era, local and state (*Länder*) governments now bear the main responsibility for financing the operating costs of bus, tram and metro services. As a temporary measure, state governments in Eastern Germany are channelling extra funding to localities so that they can provide the necessary emergency funds to cover public transport deficits. The state governments in turn are receiving temporary revenue-sharing funds from the federal government in Bonn during the transition period of rebuilding the Eastern German economy. With governments at every level in deep financial crisis in Germany, it is unlikely that public transport in Eastern Germany will be able to depend on such generous funding over the long term. As a consequence, the fare increases and service cutbacks of the past few years will almost certainly continue, causing further permanent losses of public transport customers as ever more travellers purchase cars and abandon public transport forever.

Finally, it should be noted that the entire system of financing urban public transport in Germany is now being restructured. The changes will be most significant for suburban rail services. Already, the former East German railway system has been fully integrated into the West German system, the DB. Moreover, the federal railway has been privatized (as of January 1994) and is now a corporation, although fully owned by the federal government (similar to Amtrak in the USA). The railway has been segmented into three divisions: one responsible for freight traffic, a second for passenger traffic, and a third for basic infrastructure (tracks, bridges and tunnels, power and signals, and so on). As described in the policy section which follows, the suburban rail services of the DB are scheduled to be transferred to regional public transport authorities around 1996, with financial responsibility for subsidies, fare structures and service levels devolved from the federal to the state and local level. In addition, federal aid for public transport investment projects is being consolidated into a more flexibile block grant.

Whatever these changes bring in terms of new organization, it is certain that the total burden of financing urban public transport in the unified Germany will be enormous. The German Association of Transport Systems estimates the annual subsidy requirement at about DM 22 billion (about $14 billion: Verband Deutscher Verkehrsunternehmen, or VDV, 1992).

URBAN TRANSPORT POLICIES

Corresponding to the two main categories of urban transport problems discussed in the preceding section, German transport policies can also be examined in two basic groups: those dealing with automobile use and those pertaining to public transport.

Policies for car ownership and use

The promotion of automobile ownership and use extends far back in German history. Indeed, in 1933 the Nazis set up a special programme to promote car travel. The Third Reich's Autoprogramm resulted in the world's first system of superhighways (4000 km of autobahns by 1941); the development and widespread use of the Volkswagen; and the elimination of taxes and fees on automobiles (Kunert, 1988). The Autoprogramm succeeded in sharply increasing car production, ownership and use from 1933 to 1941. The devastation of the Second World War set back car travel to negligible levels, however, as German cities and transport infrastructure throughout the country had to be almost completely rebuilt. Nevertheless, it did not take long before car use rebounded and quickly grew far beyond pre-war levels.

Similar to other countries in Western Europe and North America, West Germany pursued a policy of accommodating and facilitating car use from about 1950 to 1970. Over those 20 years, car ownership in Germany grew 19-fold, and passenger kilometres of auto travel grew 11-fold.

Basically, an attempt was made to emulate American cities by adapting German cities to serve the needs of the car (Kunert, 1988). Gasoline tax revenues were earmarked to provide revenues for extensive road construction. From 1950 to 1970, the urban road network expanded by 53 000 km (see Table 3.6). In addition, large multi-level parking garages and extensive car parks were built in virtually all German cities, further encouraging car use.

From 1970 onwards, policy towards the car shifted considerably, from accommodation to restriction and modification of automobile use. The damage inflicted by cars on both the natural environment and urban communities was becoming ever more apparent. Environmentalists, community activists and urban planners have increasingly viewed the car-

Table 3.6 Supply of roads in Western Germany, 1951–93 (in thousand km)

Year	*Interurban roads*		*Urban roads*
	Autobahn	*Total*	
1951	2.1	127.6	217
1960	2.5	135.3	227
1970	4.1	162.3	270
1980	7.3	171.5	308
1990	8.8	173.9	325
1993	9.1	174.1	331
1993*	11.0	226.8	413

*Eastern and Western Germany combined.
Source: German Ministry of Transport, *Verkehr in Zahlen* (1982–93).

dominated transport system as an important cause of the social, economic and environmental problems of German cities. In response, governments at various levels have adopted a range of measures to 'tame' the automobile. The intent is not to limit ownership, but to mitigate the harmful side effects of excessive use.

Traffic calming

Especially since 1980, most West German cities have reduced speed limits in urban residential areas to 30 km per hour and have further discouraged traffic by narrowing streets, increasing the number of curves and installing speed bumps, ornamental posts ('bollards'), concrete planters, wider pedestrian walkways and bicycle lanes. Such traffic-calming measures are aimed at reducing car use or at least making it less dangerous for pedestrians and cyclists. Speed limits and other traffic regulations within cities are strictly enforced, not only directly by police officers but also by remote cameras that monitor traffic and automatically photograph cars speeding, failing to stop or yield right of way, or violating various other regulations.

Car-free pedestrian zones

In virtually all German cities – as well as many smaller towns and villages – there is an interlocking system of streets in the old town centre and main shopping district that is almost completely off-limits to private cars

(Hajdu, 1989). Most such zones enhance pedestrian and bicycle access to the very heart of the city while keeping cars at a distance, forcing them to park in fringe car parks and garages. By contrast, public transport is allowed direct access to this central zone, although buses and trams are usually required to travel at reduced speeds to ensure the safety of pedestrians and cyclists.

Parking restrictions

Parking in German cities has become more difficult and much more expensive since 1980 (Pucher and Clorer, 1992; Topp, 1993). In most cities, the price of on-street metered parking increases considerably with proximity to the city centre. The largest cities now charge at least DM 5 per hour for parking in the centre, roughly equal to the price of a round trip by bus, tram or metro. Moreover, special parking meters have been installed to prevent long-term parking by commuters in residential neighbourhoods and, as a further disincentive, residential area parking permits are increasingly required for non-metered parking spaces.

Right of way priorities

Most German cities have constructed extensive systems of bicycle lanes and bikeways that are separated from motorized traffic. Buses benefit from reserved lanes on many streets, and both buses and trams are often given priority at traffic intersections, with lights automatically turning red for cars and green for buses and trams whenever the latter approach shared intersections. Such priority lanes and traffic signals obviously increase the average speed of public transport and make it more attractive relative to the car.

Vehicle manufacturing standards

Under the pressure of government mandates, car manufacturers have been required to produce safer, quieter and less polluting vehicles. Most of these manufacturing standards have been less stringent than in the USA and were introduced later. Nevertheless, cars in Germany and the rest of Western Europe have certainly improved since the 1970s.

Most notable in this respect are the increasingly strict emission standards for new cars since the mid-1980s. They roughly correspond to those in the USA, imposing maximum permitted levels for CO, HC and NOx (Umwelt, 1992 and 1993). To achieve those emission reductions, all new cars are now equipped with catalytic converters. Moreover, unleaded petrol is now available virtually everywhere in Germany and increasingly throughout the EU. The use of unleaded petrol is absolutely essential for the proper functioning of catalytic converters, and dramatically reduces airborne lead emissions.

At first, the West German federal government tried to encourage the installation of catalytic converters through various tax incentives such as reduced tax rates for cars with catalytic converters and for unleaded petrol. It was not able to require absolute emission standards, however, without coordinating them with all other EU member countries. Due to extreme concern for its dying forests, Germany pushed hard for uniform European standards, which were in fact adopted in 1985 and revised in 1991 and 1992 (Umwelt, 1992 and 1993). Different types and sizes of motor vehicles are subject to different deadlines and different standards but, in general, emission standards for all motor vehicles throughout the EU will be much stricter than in the past. As in the USA, the reduced emissions per kilometre driven will help offset the increase in air pollution that would otherwise result from the enormous growth in car travel.

Manufacturers have not only been required to produce cleaner cars but safer cars as well. European safety standards have been less extensive, less strict and less coordinated than in the USA. The most important safety feature is the seat belt. In combination with laws that require occupants to use them, seat belts have dramatically reduced the traffic fatality rate in Germany and in most other EU countries (Bruehning, 1993).

Regulating driving behaviour

Although Germany and other European countries have generally lagged behind the USA in regulating car manufacturers, they have been at the forefront in regulating driver behaviour. Particularly in Germany, licensing requirements, vehicle inspections, drink-driving laws, seat belt use laws, parking restrictions and urban speed limits are much more stringent and also much more strictly enforced than in the USA. Each of these regulations is intended to mitigate the harmful impacts of the car by improving the behaviour of the user. Such consumer regulation is obviously less popular than the technological fixes required of the manufacturers, but in Germany it has encountered less political opposition than in the USA, for example.

Taxes on car ownership and use

Germany imposes considerable taxes on car ownership and use. Even in 1989, before various tax increases, total road user taxes by all government levels in Germany were more than double the total public expenditures on road construction, maintenance and administration (IRF 1991). Since 1990, taxes on car use have been raised considerably. Most recently, the petrol tax was raised by DM 0.16 (about $0.10) per litre on 1 January 1994, bringing the total tax to DM 0.98 per litre (about $3 per gallon). That tax rate is roughly average for Western Europe but six times as high as in the USA.

There is also an annual motor vehicle excise tax, ranging from DM 13.20 to DM 45.50 per ccm engine size. The total tax depends on the size of the motor, whether it is petrol or diesel powered, and whether its emissions are high or low.

Finally, for the past few years, there has been considerable debate about introducing an annual permit fee for use of the autobahns. That proposal has not yet received enough political support to allow its enactment into law. All versions of the proposal foresee lowering the motor vehicle excise tax to offset the increased tax burden of the autobahn fee. Overall, therefore, it would not increase the average tax on car use but would redistribute it among various groups of users and, most important, would force payment by foreigners using the German autobahns as transit routes.

Supply of urban roadways

As shown in Table 3.6, urban road capacity has grown considerably over the past four decades. The fastest growth was from 1960 to 1970 (+19 per cent), with successive slow-downs from 1970 to 1980 (+14 per cent) and 1980 to 1990 (+6 per cent). As discussed previously, the slow-down in roadway construction reflects the new philosophy of limiting and chanelling car use rather than accommodating it. Moreover, fiscal constraints and opposition from environmental and community groups have also limited road expansion.

Financing for urban roads comes from all government levels. The more important a road is for long-distance traffic (major highways, bypasses, connectors), the more state and federal aid is provided. Streets with exclusively local traffic are financed primarily by the cities themselves.

The federal government dedicates a proportion of petrol tax revenues specifically for urban transport infrastructure projects, including roads. In Eastern Germany, the federal share of financing was 90 per cent in 1992, 85 per cent in 1993, and 80 per cent in 1994; in Western Germany, it is 75 per cent. Both in the East and the West, state and local governments must provide the rest of the funding required. Roughly 60 per cent of total federal funding for urban transport infrastructure goes to public transport and 40 per cent to roads. Prior to 1992, states had little discretion in shifting funds among modes but, since then, the law has been amended to allow the states to determine on their own how 80 per cent of the federal urban transport subsidies will be spent. The states individually allocate the funds between roads and public transport, and also decide which specific projects will be financed (Fromm, 1992).

Policies towards public transport

The past and current systems of financing and organizing public transport in Germany have already been discussed in the previous section on problems in public transport. This section will, therefore, focus on emerging developments.

Suburban rail

Suburban rail services are by far the most problematical of the public transport services in Germany and are also the most burdensome in terms of subsidy needs. Thus, most of the impending changes in organization and finance pertain to suburban rail.

For decades, the federal government in Germany was responsible for covering all capital and operating costs of passenger rail services of the DB (and of the Deutsche Reichsbahn from East Germany since 1990). That included all the short-distance suburban rail services within urban areas as well. As it turned out, it is precisely those local services which have been by far the most unprofitable of all rail passenger services and which have required the largest subsidies: about DM 9 billion in 1993, or roughly half the total subsidy to urban public transport in Germany. Since suburban rail only accounts for 15 per cent of all urban public transport passenger trips and for 31 per cent of all passenger kilometres, that means that it requires about four times as much subsidy per passenger trip and about twice as much subsidy per passenger kilometre as other types of public transport.

Part of the problem was the DB itself, which for decades has been overstaffed, inefficient, unresponsive to consumer needs, completely unionized, too centralized, too bureaucratic and highly politicized. Decisions on fares and service levels have been based more on politics than on sound economic principles. To solve some of those problems, the DB was turned into a semi-private corporation on January 1994 and renamed the Deutsche Bahn. The new corporate form is intended to insulate the railway somewhat from political interference and allow it to become more productive and more sensitive to market incentives. The segmentation of the railway into three divisions – freight, passenger and infrastructure – is also intended to make the railway more manageable and to separate out distinctly different types of functions and thus increase efficiency.

In addition, the local, suburban rail services of the railway are being shed by the Deutsche Bahn altogether (VDV, 1992). It is felt that they are local or regional in nature and would be most appropriately planned, organized, provided and financed at the local and regional level. Thus, in the coming years, suburban rail services will be devolved to regional public transport authorities in each metropolitan area, much as they are in the USA. In return for accepting the huge additional subsidy burden required by suburban rail, the states are receiving additional funds from the federal government in Bonn. It is not yet certain exactly how all of this is going to work out, but the federal and state governments have definitely agreed on the reorganization plan (at least in principle).

Metro and tram systems

Aside from suburban rail, there will be no fundamental changes in German public transport. Due to financial constraints, no new heavy rail metro systems are planned in any German cities. Most of the emphasis now is on integrating the two formerly separate systems in Berlin. Moreover, many German cities are either improving or extending their tramway/light rail systems. Already there are 56 such systems in operation in Germany (half of them in Eastern Germany). The current plans envisage investing the limited subsidy funds in moderately expensive light rail systems instead of extremely expensive metro systems. Trams in West Germany had already been considerably modernized and expanded during the 1980s. That modernization and expansion will probably be continued during the 1990s, but the focus now is on modernizing and upgrading the already extensive tram systems in Eastern Germany. Unlike West Germany, where most tram systems were replaced with buses, trams were always the backbone of public transport in virtually all large and moderate-sized East German cities. Nevertheless, both tram infrastructure and vehicles in the East are in desperate need of rehabilitation and modernization, which are currently under way.

Financing

The basic federal law on aid for urban transport (*Das Gemeindeverkehrsfinanzierungsgesetz*) was amended in 1992, effectively converting 80 per cent of the federal subsidy into a block grant to be allocated at the discretion of the states between public transport and roads, as well as among specific alternative projects (Fromm, 1992). The federal matching percentage for the block grants is generally 75 per cent, with state and local governments providing the rest. In the new Eastern German states, the matching percentage was 90 per cent in 1992, 85 per cent in 1993, and 80 per cent in 1994. The remaining 20 per cent of federal urban transport funding is exclusively for large rail-oriented public transport investments. The federal minister of transport has full control over project selection for that portion of the subsidy funding, which is subject to a 60 per cent federal match and a 40 per cent share from state and local governments. The change in the federal transport subsidy law greatly decentralizes decision-making in project selection and conforms to the trend towards devolution of government responsibility in urban transport throughout Europe and North America.

Regional transport organizations

West Germany has been at the forefront of regional coordination of public transport services in urban areas. Starting with Hamburg in 1967, an increasing number of German cities have established various organizational forms of regional cooperation and integration (German Ministry of Transport, 1988d). By 1990, virtually all West German cities had integrated their public transport systems. The two most important organizational forms are the *Verkehrsverbund* and the *Verkehrsgemeinschaft*. Both types of public transport agencies enable fully integrated route networks, timetables and fare structures. From the perspective of the passenger, it is as if only one firm were providing all public transport services within each German metropolitan area. The same ticket or monthly pass can be used for any mode of public transport in any part of the region. Route maps, timetables and service standards are uniform. Transfers among modes and routes are easier both due to physical coordination of services (spatially and temporally) and due to the zone-based unification of the region-wide fare structure, which allows passengers free choice of modes and routes (including suburban rail services).

Regional coordination of public transport services has greatly enhanced the quality of public transport in Germany. In most cities, coordination has achieved increased use, or at least a slow-down in loss of customers. Nevertheless, it has not succeeded in reducing costs or subsidy needs (German Ministry of Transport, 1988d). Indeed, the total operating subsidy for urban public transport in Germany grew especially fast during the late 1960s and the 1970s, when most regional transport agencies were formed. Some of that subsidy increase was due to service expansion and fare discounts, and it is unlikely that regional coordination in itself caused deficit growth.

CONCLUSIONS

The period of rapid expansion in providing transport infrastructure and services is past. At least during the decade of the 1990s, there are no prospects for large new investments either in road or public transport projects. Financial constraints at every government level set stringent limits on any new spending. Most of the emphasis now is on improving the existing transport systems, including modernization, rehabilitation, rationalization, better integration and coordination, and increased productivity. Instead of expanding transport capacity, a range of transport management measures have been implemented to ration the existing capacity among users and to mitigate the adverse social and environmental impacts of car use. Reserved lanes, parking and speed restrictions, traffic calming, pedestrian zones and various pricing incentives are increasingly being used to alter the behaviour of urban travellers in ways that reduce the strain on the transport system and the environment.

Perhaps the most striking development over the past 10 years is the acknowledgement of the importance of non-motorized modes of urban transport. In Germany as in other European countries, roughly 40 per cent of total urban travel is either by foot or by bicycle, both of which are extremely energy efficient and environmentally friendly modes. More than ever before, German cities are investing considerable effort and ingenuity in improving facilities for pedestrians and cyclists. Wider pedestrian walkways, more extensive networks of walkways, overpasses and underpasses, bike lanes and bike paths, bicycle parking areas and even garages, priority traffic signalling, pedestrian zones and better coordination of walking and cycling with public transport: all those measures are helping to encourage more walking and cycling and less reliance on the car. Only the Netherlands, and perhaps Denmark, exceed Germany in such efforts to promote non-motorized modes of urban transport. Since they are the least expensive – both from a private and from a public point of view – and because they cause virtually no adverse environmental or community impacts, walking and cycling may well be the trend modes of the 1990s.

4 France: The Impossibility of Accommodating the Car and Public Transport in Transport Policies

France is well known for the quality of its public transport technology and in particular for the 'Train à Grande Vitesse' or TGV (High Speed Train) and the 'Véhicule Automatique Léger' or VAL (Automatic Light Rail) which are the twin jewels in its crown. But these success stories cannot hide the fact that France is also a country where the automobile is cherished by both the population and the political authorities. In the 1950s and 1960s, central government launched a vast programme of investment in roads and the automobile, the car industry being considered a key sector in the country's reconstruction and economic growth. The public authorities did not, however, invest in public transport in the same way, or at least not until the early 1970s. In this respect, France differs from Northern European countries but is very similar to those in Southern Europe. Consequently, French urban areas are increasingly plagued by traffic problems although severe road congestion and environmental problems only exist in the largest areas.

However, the French people's love of the car has not reduced awareness of the importance of public transport. On the contrary, since the 1970s French urban transport has been experiencing a renewal which many see as the result of its unique and paradoxical regime. Although France is a very centralized state with a strong philosophy and policy of state intervention, this is not reflected in urban public transport. Indeed, this sector has been almost completely decentralized, both politically and economically, and the private sector occupies a larger place than in any other European country, albeit controlled by local governments. Such a system, midway between full public ownership and total deregulation and privatization, is now being considered with interest by most other European countries. However, although urban public transport has shown significant positive changes in the last decade, its present situation is not particularly healthy with some decrease in usage and a persistent lack of funding. Indeed, urban public transport in France is once again facing a severe crisis as are transport policies, the 1980s having promised much but achieved almost nothing in terms of comprehensive transport policies. These are issues which will be developed later, but let us first examine the context and overall situation of urban transport today.

URBAN SPATIAL STRUCTURE

The urbanization of France is recent when compared to that of the USA or north-western European countries, but its present urbanization rate is average for Western Europe. Today, about 75 per cent of the French population, that is to say 41 million people, live in urban areas, half of them in the 30 urban areas which have over 200 000 people each (see Table 4.1). In 1990, this represented a total of about 21.5 million people. Urban growth was rapid and strong in the 1960s and early 1970s but in the biggest cities, (those with over 200 000 inhabitants), there was a clear slow-down in the intercensus period of 1975–82, followed by a slight upswing in the period between 1982 and 1990 (see Table 4.2). Urban growth took place first in the suburban areas which today house 43 per cent of the French urban population (17.6 million people) and which continue to represent an important part of urban growth. Since the early 1970s, however, the most dramatic increases have taken place on the outskirts of the traditional urban areas, in places which are still rural in character and where today about 9.7 million people live. During the period from 1975 to 1990, suburban areas grew by about 0.9 per cent per year, compared to 1.6 per cent a year in the outer suburbs. At the same time, the population of city centres remained unchanged, and even increased slightly in the biggest areas.

Table 4.1 Population evolution by type of urban area, 1975–90

	Population in 1000s			*Annual rate of change (%)*	
	1975	*1982*	*1990*	*1975–82*	*1982–90*
France	52 656	54 335	56 614	+0.52	+0.59
City centres	23 565	23 413	23 540	–0.09	+0.07
Suburbs	15 455	16 446	17 597	+0.89	+0.97
Rural peripheries	7 827	8 746	9 687	+1.87	+1.47
Rural areas	5 809	5 731	5 791	–0.23	+0.15

Source: INSEE, population censuses.

Table 4.2 Population evolution in the 10 largest urban areas, 1982–90 (in 1000s and %)

Urban area	*Population (1982)*	*Population (1990)*	*% annual change*
Paris	8 924	9 319	+0.55
Lyon	1 220	1 262	+0.43
Marseille	1 249	1 231	–0.18
Lille	945	959	+0.18
Bordeaux	647	696	+0.94
Toulouse	569	650	+1.80
Nice	477	517	+0.10
Nantes	467	496	+0 77
Toulon	410	438	+0.85
Grenoble	396	405	+0.28

Source: French national censuses (1982 and 1990).

Population growth has therefore occurred in places which are more and more distant from the traditional urban cores and which are very often not contiguous with existing urbanized areas. This spatial extension of cities has significantly reduced their density. In 1982, the average density of French urban areas was 494 people per sq. km. Today, it is 467. In the peripheral areas of the largest cities, density decreased considerably in the 1980s, falling from 759 to 715 (Geffrin and Muller, 1993). The density of urban settlements is therefore falling and they are encompassing increasingly large territories, sometimes separated from the 'traditional' urban area by spaces which are rural in character.

The decentralization of the population has been accompanied by a decentralization of activities, although city centres still maintain their primordial importance in terms of employment. To a large extent, French urban areas remain dependent upon their core even though some important secondary poles have emerged on their peripheries, a trend which can be seen in the spatial evolution of urban trips.

TRENDS IN URBAN TRANSPORT: THE DOMINANCE OF THE CAR AND THE EROSION OF PUBLIC TRANSPORT

There is no national data on urban mobility. Existing data is a compilation of the results of household surveys carried out in several urban areas at various times by the Ministry of Transport. Most of the information found below is therefore derived from these Ministry of Transport compilations (notably Guidez, 1990 and Geffrin and Muller, 1993).

On average, every urban resident makes about 3.3 trips per day. This number has remained more or less stable over the last decade for urban areas as a whole, and has fallen slightly in the case of the largest cities. As in other western countries, trip purposes have evolved significantly. Whereas 15–20 years ago the journey to work represented a large proportion (roughly 50 per cent) of the total trips made by the urban population, it now accounts for only a small proportion, 19 per cent in Grenoble (1992) for instance, and 24 per cent in Toulouse (1990). More and more trips are now made for 'other' purposes (leisure, shopping, and so on) and current figures for the Grenoble and Toulouse areas are 61 per cent and 59 per cent respectively. The length of the average trip seems to have increased considerably, although we do not have any comprehensive comparative data on all trip purposes. If we consider the journey to work, its average length more or less doubled between 1975 and 1990, from 7.4 to 14 km. This is largely due to the considerable increase in the number of commuters over the last 15 years (in 1975, 60 per cent of the labour force lived and worked in the same town; the 1990 figure is only 45 per cent). However, this has had no effect on the amount of time spent travelling to work since it has remained relatively stable over the last 15 years, mainly as a result of improvements in public transport services and road traffic conditions.

The spatial extension of urban areas and changes in the localization of residences and activities have played a major role in the evolution of trip patterns. Today, the vast majority of motorized trips are no longer linked to the urban core. About 75 per cent of all trips made in urban areas are between peripheries (75 per cent in Grenoble, 66 per cent in Paris) and it is these that have shown the greatest increase (+6 per cent in Grenoble, +12 per cent in Paris over the last decade).

The car remains dominant for all journeys. In 1990, modal split in the densest areas (those with over 3300 people per sq. km) was as follows: car 54 per cent; public transport 12 per cent; two wheelers 4 per cent; and walking 30 per cent. If one considers motorized modes alone, the car accounts for about 75 per cent of all trips. In France, car ownership increases by 4 per cent each year. The number of private cars per 1000 people increased from 354 to 420 between 1980 and 1992. This increase in car ownership means that in 1990, 77 per cent of households possessed a car and 27 per cent owned at least two cars. In 1992, the fleet comprised 24 million vehicles. In 1980, daily car mileage in all urban areas with over 20 000 people was 141 million km. Ten years later, it was almost 200. The dominance of the car can be seen everywhere except in city centres (in Paris, 56 per cent of households do not own a car!). While public transport use remains virtually stable, car use is increasing by between 1 per cent and 2.5 per cent per year, depending on the urban area in question. Even in the Ile-de-France region (the Paris area, or RIF) where its importance is less than in other large cities, it increased from 61 to 66 per cent of all trips between 1983 and 1991. On the other hand, the use of two wheelers is gradually declining, as can be seen from Table 4.3.

The dominance of the car is becoming more and more important in the growing market of suburb-to-suburb trips. In Grenoble, for instance, when public transport gains one trip, the car gains 10 and the two wheeler loses 7.5. In 1992, the car accounted for 78 per cent of trips within the Grenoble periphery. In Bordeaux, in 1990, it accounted for 83 per cent, in Nantes for 79 per cent and in Rennes, 84 per cent.

With an average of 12 per cent of all motorized trips in 1990, public transport remains significant, all the more so in the largest urban areas (RIF 30 per cent; Lyon 25 per cent; Marseille 22 per cent; Nantes 17 per cent; Toulouse 13 per cent). But it is worthwhile analysing this situation over a long period. In 1975, the average French urban resident made about 58 trips a year by public transport. In 1980, the figure was considerably higher and stood at 74 trips. In 1991, the number of trips per inhabitant had risen to 94. This evolution is not linear, however. As shown in Table 4.4, the end of the 1970s was a 'good' period for public transport. Since then, the growth of public transport use has been less rapid, and showed its first decrease for about 20 years in 1991 and 1992. It is too soon to tell whether the 1.8 per cent increase in 1993 can be considered as the beginning of a new period of growth. The RIF is illustrative of this situation,

Table 4.3 Change in modal split in Grenoble and Bordeaux urban areas (%)

Year		*Car*	*Public transport*	*Two wheeler*
Grenoble	1978	65	17	18
	1985	75	17	8
	1992	75	19	6
Bordeaux	1978	69	15	16
	1990	79	12	9

Sources: Household surveys, corresponding years.

Table 4.4 Annual changes (%) in public transport use, 1975–93 (RIF excluded)

1975–80	*1980–85*	*1985–90*	*1990–92*	*1993*
+ 5	+ 2.3	+ 2	–0.3	+ 1.8

Sources: CETUR, Données et analyses sur les transports urbains, Ministère des Transports; UTP (1994).

with public transport modal split decreasing from 32 per cent to 30 per cent between 1983 and 1991.

The evolution of urban modal split in France since the 1970s can thus be summarized as an increase in the dominance of the car coupled with the erosion of public transport. It may be partially explained by the development of road infrastructure during the same period.

The increase in car ownership over the last decades has made it necessary for new roads to be built in urban areas. Although a great deal of road construction was undertaken in the largest cities in the 1960s and 1970s, the situation was rather different in medium-sized cities where traffic conditions were considered less dramatic. This situation was remedied in the years between 1980 and 1990 when by-passes, ring-roads and urban expressways were finally built in most of these urban areas. Almost 50 per cent of the projected investment in the national road network forecast in the 1989–93 Plan is being used to build new roads or enlarge others in urban areas with a population of over 20 000. In the largest cities, the increase in traffic combined with urban sprawl has necessitated the extension of urban expressways in the suburbs (a 160 per cent increase in road mileage between 1970 and 1990: Orfeuil, 1993) and in areas like the RIF new ring-roads, further out in the periphery, were planned and are now being built. With the relative exception of the largest areas and in particular Paris, French cities now have a fairly good road network which explains why the amount of time spent in commuting trips has remained constant over the last 15 years, even though the average distance between home and work has almost doubled.

Improvements in the urban road network have been paralleled by the betterment of public transport infrastructure. In the 'good' period of public transport, supply increased by about 7 per cent a year in km per inhabitant, and by 9 per cent in seat km (Lefèvre and Offner, 1990). This growth has continued, but less rapidly than in the 1980s and now stands at about 2 per cent a year (UTP, 1994). France, like certain other industrialized countries, eliminated its trams in the 1950s and 1960s. In the early 1970s, Paris was the only urban area to possess a rail system. The 6th National Plan (1971–75) prescribed the building of three new metro lines in the three major cities outside Paris (Lille, Lyon and Marseille) and by the end of the 1970s, Lyon and Marseille had their own metro, followed in 1983 by Lille with the first fully automatic metro line in the world (VAL). Since then, the extension of these networks and the building of light rail systems in smaller cities has become one of the country's major public transport issues. So far, five new light rail systems have been brought into operation (Nantes in 1985, Grenoble in 1987, Toulouse in 1993, Strasbourg and Rouen in 1994), and many more are being planned and built.

THE URBAN TRANSPORT CRISIS

The evolution of urban travel patterns in response to urban development over the last three decades has played a major role in the emergence of the urban transport crisis. Its impact on urban society as a whole and on urban transport in particular has been tremendous, although it is difficult to find aggregate data on this for the entire country.

As far as the impact on urban society as a whole is concerned, the dominance of the car is having a marked effect in at least four domains: air pollution, energy consumption, safety and social exclusion. It is now well known that the transport sector, and in particular road traffic, is largely responsible for air pollution. In France, road transport accounts for 90 per cent of all pollution from transport and within urban areas, private cars are responsible for about 50 per cent of all CO, and 65 per cent of all HC, emissions (Morchoine, 1993). Any increase in car use is therefore likely to entail a worsening of air pollution in cities. However, the latest comparative data shows that between 1980 and 1988 total emissions declined by 9 per cent (CETUR, 1994). This does not, however, mean that French cities do not experience severe air pollution problems, but at present these are episodic: that is, they occur in a few large urban areas (Paris, Lyon, Strasbourg, Rennes) and in specific circumstances only (for example, absence of wind and rain to clean up the atmosphere). In the field of energy consumption, the dominance of the car augments the country's energy dependence since most petrol is imported. In 1980, the transport sector was responsible for 40 per cent of all petrol imports. Ten years later the figure had risen to 60 per cent, and 80 per cent of this was for road transport. Urban transport (mainly cars) is responsible for about 40 per cent of all transport consumption and is increasing four times faster than non-urban transport. Urban transport consumes about 11 million British thermal units a year, 10 million of which are used by the car. Unless action is taken, air pollution and energy consumption due to the car will increase in the coming years. The situation is all the more dramatic since urban congestion will play a major role in this growth. Studies have shown (Morchoine, 1993) that energy consumption increases three-fold in heavy traffic, while HC and CO emissions increase almost four-fold.

Road safety is also becoming an important issue. France is well known to be one of the most dangerous countries in Europe, since between one-fifth and one-sixth of total EU fatalities occur there. In 1972, a peak was reached, with almost 17 000 people killed and nearly 390 000 seriously injured on the roads, for a population of only 50 million and 16 million registered vehicles. In 1987, the number of deaths on the road fell below the symbolic threshold of 10 000 for the first time. Indeed, between 1972 and 1992, the number of accidents decreased by 30 per cent and the number of fatalities by 23 per cent. Urban traffic is largely responsible for the high number of accidents with 72 per cent of road accidents occurring in urban areas. Around 17 per cent of corporal accidents involve pedestrians, nearly all of them (93.5 per cent) in urban areas. However, only 37 per cent of fatalities occur in cities because, on average, urban accidents are less serious than those occurring in country areas. In 1991, there were about 3500 fatalities in urban areas against about 5600 in 1970. This decrease is all the more significant since in the same period the population grew by about 13 per cent and car ownership by about 30 per cent (not to mention the drastic increase in car use particularly in urban areas).

Current forms of urbanization have produced sociospatial segregation, the social impacts of which are becoming increasingly severe. It is now the case that urban France is experiencing a sociospatial dualism which can be seen in the mobility patterns of its residents. French cities have, like any other urban areas, neighbourhoods with a high concentration of poor people and ethnic minorities. Although some of these neighbourhoods are located within the city centres, most of them are to be found in the suburbs. These areas, which display all the classic elements and statistics of poverty and social exclusion, are nevertheless generally fairly well served by public transport (CETUR, 1993a), albeit with some problems of accessibility, as well as problems which may also be found in other neighbourhoods. The latter include safety problems for the public transport operators which may end in the elimination of some services considered to exist essentially for the journey-to-work, which means an absence of services after 8pm, and so on. It is generally acknowledged that people in these areas have to rely on public transport to a greater extent than others. Although mobility in these neighbourhoods is not very well documented, studies have recently been undertaken (CETUR, 1993a) which show that car ownership in these areas is significantly lower than in the rest of the city. For instance in Lille, while the average ratio for the city is 0.92 car per household, it is only 0.64 in the 'socially disadvantaged' area. In Toulouse, the figures are respectively 1.21 and 1.03. This obviously has a direct impact on the residents' mobility. In Grenoble, for example, whilst the average person makes 3.6 trips a day, the figure is only 3.3 for the poorer neighbourhoods and the same applies in other urban areas like Dunkirk (3.9 and 3.7 respectively) and Lille (3.8 and 3.5). The difference in mobility between a resident of a poor neighbourhood and that of the average citizen makes the improvement of public transport services in these areas all the more urgent, which means an increase in the frequency of services, an extension of service periods, and above all, the creation of suburb-to-suburb lines because the 'radiality' of the existing networks plays a major role in the 'feeling of exclusion' experienced by residents of these neighbourhoods.

The effects of car dominance on the transport sector itself are also considerable. Some, like urban congestion or parking problems, are clearly visible in daily life; others, such as the difficulties of financing infrastructure or public transport, are less immediately apparent but nonetheless crucial. It is difficult to know the exact situation regarding urban congestion in France. One has to rely on fragmented data or studies on specific urban areas which may not be relevant to the average French city and are not easily comparable. Urban congestion is considered one of the most important issues in large urban areas, like the RIF. In 1989, the cost of urban congestion in the Paris area was estimated at about FF 2 billion. Traffic jams on urban expressways in the RIF (in hours/km) increased by 600 per cent between 1974 and 1989 since when the annual growth, expressed in hours/km, has been 17 per cent. Although this seems specific to the largest areas (RIF, Lyon, Lille and Marseille), it is now believed that this type of situation is occurring in areas with under 100 000 inhabitants. The same applies to parking problems which are more critical in these same cities. In Paris, for example, the lack of parking spaces means that two out of three residents have no option other than to use on-street parking. However, with the exception of the largest urban areas and some city centres, it must be admitted that the current situation is really rather good; on average, traffic and parking conditions are fair, and French cities are not yet experiencing the problems encountered in the USA or in very dense European areas like the Randstad or the Ruhr.

The need to accommodate a dramatic increase in car traffic (+4.4 per cent for urban expressways and national highways in 1988), which seems to be structural, has necessitated the development of new projects for road construction and other facilities (urban tunnels, bridges, and so on). All these facilities have been financed by the General Treasury and local government budgets. However, the building of roads in urban areas has proved difficult since there has been insufficient public money to fund all the projects and construction has often been delayed or postponed (CETUR, 1990a). This situation also applies to public transport. About 30 urban areas currently have plans to build reserved-track public transport systems, busways or light railways, but so far financial difficulties have meant that only a few have been able to start building.

It is obvious that the dominance of the car has also had a major impact on public transport itself. We have already seen that public transport networks have experienced a period of stagnation, and even decline, in usage. This decline is largely due to the spatial patterns of recent forms of urbanization, characterized by low density and dispersion, a situation which undoubtedly favours the car. Urban congestion, which has meant a decline in commercial speed and the quality of services, making them less reliable, has also played a major role in lessening the attraction of public transport. All this has had a definite impact on the financial situation of most networks.

Between 1975 and 1985, the development of urban public transport use and supply was achieved, but resulted in a drastic decrease in the fare recovery ratio. To halt this decline, productivity measures were introduced. These were partially successful and, at the end of the 1980s, the fare recovery ratio increased slightly (see Table 4.5 and Lefèvre and Offner, 1990). This no longer seems to be true, since the last available figures (UTP, 1994) show a 1 per cent annual decrease in this ratio since

Table 4.5 Fare recovery ratio in public transport, 1975–93 (%)

Area	*1975*	*1980*	*1985*	*1990*	*1993*
Province*	77	54	49	54	55
RIF	34	36	33	38	

*All urban areas, RIF excluded.
Source: CETUR, UTP and Direction Régional de l'Equipement d'Ile-de-France.

1991, which amounted to about 55 per cent of total operating costs in 1993 for urban public transport outside the Paris urban area (RIF).

Consequently, public transport has required more and more subsidies, which have been taken from the local transport tax and even local government general budgets (in 1980, the transport tax represented about FF5 billion at current values; in 1993, about FF12 billion at current values, 57 per cent being levied in the Paris urban area only).

POLICIES DESIGNED TO DEAL WITH THESE ISSUES

The crisis in urban transport is first and foremost the result of *laissez-faire* and of the failure of policies to deal with the issues mentioned above. Urban sprawl and the decline of centrality and consequent peripheral mobility, the extension of catchment areas far beyond the urban cores and spatial segregation in the suburbs are societal issues which the authorities have only just started to tackle. Urban congestion, safety, energy consumption and parking problems have been partially diagnosed, but remedies have yet to be found and the financing crisis in public transport and infrastructure building has yet to be dealt with. This is not to say that no action has been taken or that no policies have been implemented, but there has been a lack of coherence, comprehensiveness and sometimes of awareness which has prevented the situation from being dealt with successfully.

First, the dominance of the car may be interpreted as the logical result of (1) the spatial patterns of urbanization and their consequent impact on mobility, and (2) policies which have continued to favour the car either by *laissez-faire* or by making this transport mode still more attractive. In the 1950s and 1960s, the car ruled supreme, as in most countries. It was a period of intense activity as far as urban road projects were concerned and the aim was clearly to open up the city to the car (Napoléon and Ziv, 1981). Although this is no longer true, from the early 1970s onwards the car continued to be cherished by the public and elected officials, again as in most countries. Even today, it is not easy to find urban areas where the car has been seriously constrained by regulations and policies. In the 1980s, many surveys and polls showed that public opinion in favour of the car was waning, but this is no longer true and proposed measures to restrain automobile use in urban areas found less support in 1992 than they had a few years previously. This change will play an important role in the shaping of urban public transport policies in the coming years.

Apart from a societal consensus not to make any serious attempt to dislodge the car from the position it has attained over the last decades, the fact that the cost of car ownership and use have continuously declined over the last decades has been a strong factor in maintaining, and even increasing, its use. In this context, the continuous and substantial decline of alternative modes such as walking and two wheelers (for reasons of safety) is easy to understand.

We will now turn to the policies that have been implemented in the last two decades to cope with the principal issues discussed above. For the sake of clarity, we will distinguish between three types of policies: (1) those intended to preserve the human and physical environment from automobile nuisance (that is, those which tackle the issues of pollution, safety and energy saving); (2) those which aim to solve the problems caused by car movement (that is, congestion and parking problems); and (3) those which address the urban transport question in a more comprehensive way, notably by attempting to create a more balanced transport system by pointing towards the development of public transport and other modes.

Preserving the human and physical environment from the automobile: policies concerning safety, the fight against pollution, and energy saving

With regard to the preservation of the human and physical environment, France has been comparatively very backward, particularly in relation to northern European countries (Scandinavia, the Netherlands, Germany and Great Britain). Safety became a national issue in 1972–73 when the number of road fatalities reached its peak, but significant measures regarding urban areas came much later. The first 'traffic calming areas' were introduced experimentally in 1977 in medium-sized cities (Alençon, Chambéry) but at national level, the first comprehensive policies were introduced at the beginning of the 1980s, after the enactment of the LOTI (the 1982 law on domestic transport) and most of the decentralization laws.

Although not specifically dealing with urban transport, the LOTI launched some innovations such as the 'right to transport' and the Urban Travel Plans (Plans *de Déplacements Urbains*) or PDUs. In the PDUs, the concept of urban travel is central and integrates all the existing modes in transport planning, clearly specifying the inclusion of so-called alternative modes such as walking and cycling, in the hope of increasing their use. Safety was considered the most important question if these modes were to be developed. Some urban areas, notably Lorient, Rennes and Annecy, took the PDU seriously and implemented policies designed to encourage walking and cycling, the creation of urban pathways in city centres being one example.

Gradually, central government launched three contractual programmes with local governments, the aim of which was to reduce road accidents and to improve the quality of life within urban areas (Loiseau, 1989).

REACT (*REAGIR*) was the first of these. Its main purpose was to obtain qualitative data on road accidents. Between 1983 and 1987, more than 10 000 technical investigations on serious accidents were undertaken, one-third in urban areas, in order to establish their exact cause and suggest an appropriate solution. A second programme, 'Goal: Minus 10 per cent' (*Objectif* –10 per cent) was enacted at the same time, with the aim of creating local policies on road safety which would bring about a 10 per cent reduction in accidents. Measures such as the construction of 'sleeping policemen', road-user education and speed reduction were implemented in 90 per cent of areas with over 50 000 inhabitants. Between 1983 and 1986, the Ministry of Transport gave local governments about FF100 million for this programme.

The third programme, 'Safer City, neighbourhoods without accidents' (*Ville plus sûre, quartiers sans accidents*) was launched in 1984 with the aim of reconciling road traffic and the quality of urban life. This programme focused on major thoroughfares, since 70 per cent of urban accidents take place on these roads. It was stopped in 1987, when the government considered that local governments should take the lead. An assessment of this programme (CETUR, 1990b), which by 1987 concerned about 60 urban areas, shows that the annual number of corporal accidents dropped by 60 per cent in the test zones. In November 1990, the government, after publishing a White Paper on road safety, decided to lower the maximum speed limit in urban areas to 50 km/hour and to allow local governments to impose a 30 km/hour speed limit on specific sections of their road network. So far, these measures have proved successful, bringing about a 10 per cent reduction in fatalities and a 15 per cent reduction in severe injuries in 1991 (CETUR, 1992b).

These programmes and measures all prove that the French government has not been inactive in the domain of road safety. It is, however, an issue which is the direct concern of local authorities, whose responsibility it is to use the legal tools in their possession to enforce measures.

Although the question of urban pollution has been on the political agenda all the more prominently since the emergence of a sizeable environmentalist movement at the last local elections (in spite of a majority rule electoral system, some environmentalists have been elected in municipal and regional councils, as in the Paris area), it has not really been given serious consideration. It is not even mentioned, for example, in the report of the 11th National Plan (Commissariat Général du Plan, 1993) regarding cities, although it was considered in various other reports (Transport Destination 2002, 1992a; Commissariat Général du Plan, 1992). However, certain measures have been approved and others implemented. The latter include, to name but a few, the opportunity to use unleaded petrol (only since 1990), the development of electric cars, compulsory pollution tests on old cars, and so on. Some urban areas such as Rennes have already shown interest in implementing more comprehensive measures to deal with urban pollution from cars, but so far it seems likely that most French cities will rely on technology rather than on any more drastic measures to prevent this. The same also applies to the limitation of energy consumption, which is not taken seriously as a political issue.

Solving the difficulties created by the movement of cars: reducing congestion and making parking easier

Urban congestion in France is still limited to the largest areas – that is to say, the RIF, Lille, Lyon and Marseille – but about 25 per cent of the total population live in these cities and, so far, no truly effective policies have been developed to tackle it. Traffic regulation was, of course, introduced at the beginning of the 1970s with the Traffic Management Plans (*Plans de circulation*). From the first technological innovation (the 'Gertrude' system in Bordeaux in 1972) to the systems now existing in most French cities ('Surf' in Paris, 'Sirius' in the RIF, and the new 'Gertrude' systems in many other urban areas), traffic regulation has taken many forms including traffic light management and user information systems but, to date, although it has helped to achieve freer flowing traffic on many roads, it has been unable to eliminate traffic congestion. Other measures are therefore now under consideration, from the construction of urban expressways to road-pricing.

Urban expressways are financed jointly by both state and local governments. In 1986, the newly elected rightist administration was firmly in favour of the involvement of the private sector in urban transport infrastructure and many projects were therefore undertaken (the A14 Expressway in the RIF; suburban expressways in Toulouse; underground expressways in Marseille, Paris and Grenoble). With the exception of Marseille where an underground expressway opened in 1993, no urban expressways have been constructed with the assistance of the private sector ... yet. The same situation applies to road-pricing. There has been much talk about this, but so far no concrete measures have been implemented.

Less costly measures such as traffic calming or other restrictions on the use of cars have not been extensively employed, and when implemented have not been seriously enforced, showing that preserving the dominance of the car is still the order of the day. This last remark also applies to parking issues. Although it is now fully accepted that parking should be paid for by car users (paid on-street parking was introduced in the late 1960s and is now widely used), it is also acknowledged that parking supply must meet parking demand. Parking problems principally arise in city centres where a wide range of car parking (residential, work, leisure, economic activities) must be accommodated. Paid on-street parking, despite being extended to the city centre as a whole, has been unable to cope with parking demand. Consequently, local governments have decided to develop public parking structures in central areas (for instance, the Paris authorities have approved the construction of 5000 spaces annually) and private car parks are also being built. This policy is not really controversial. Now and then environmentalists, experts and some governmental officials do criticize these measures on the grounds that rather than solving the parking problem they will increase it, because building car parks in city centres will make these areas all the more attractive to motorists. Most local government officials, however, do not think this way. In France, the motorist is still an important voter and, surprisingly, public opinion was more in favour of the car in 1993 than it was a decade earlier.

Achieving a more balanced transport system

Policies directed towards a more balanced transport system (that is, the revitalization of public transport) were launched in the early 1970s, after the Congress of Tours in 1970. In 1971, the *Versement Transport* (VT) was created in the Paris area. This was originally a payroll tax that local governments could levy on companies with more than nine employees, located within the Urban Transport Perimeter (*Périmètre de Transport Urbain*, or PTU: see below). It was to be used to make up operating deficits due to fare reductions for the journey to work and to finance infrastructures provided they were related to the journey to work.

The VT, which is unique in the world in its present national form, was then extended to other urban areas, first in 1973 for areas with over 300 000 people, then in 1974 for those with over 100 000, then again in 1982 for those with over 30 000; the population threshold was finally reduced to 20 000 in 1992. The general idea of the VT was that public transport development would be supported by the economic sector, thus consolidating and extending the employment catchment area. To achieve this objective, the original idea was to create a PTU. A PTU is theoretically the urban territory served by public transport. It comprises several towns which must cooperate to form a transport authority. This authority may then levy the transport tax provided it reaches the population threshold imposed by law. The various towns located within an urban area are thus incited to cooperate. This proved successful since all French urban areas now have a public transport authority whose territory (the PTU) covers most of the contiguous urban area.

In enacting the VT laws and the idea of the PTU, the French government established a clear separation between urban and non-urban transport. Since then, urban public transport has been given clear privileges: a specific and substantial tax in terms of revenue, and also a specific territory (the PTU) within which not only local government policies but also state policies in favour of public transport have been implemented (Lefèvre and Offner, 1990); meanwhile, outside these urban areas, public transport has continued to deteriorate. Whatever the drawbacks of such a policy, it is nevertheless obvious that these measures allowed public transport to develop in urban areas in a period of great turmoil and fundamental choices for public transport.

In the 1960s, competition from the car proved fatal to public transport. Private companies could no longer survive without public assistance and, as shown in Chapter 2, France was no exception. At the time, local governments could not be, and did not want to be, directly involved in the financing of public transport, and the state did not favour explicit subsidies. In a period of steady economic growth, the involvement of the economic sector was considered suitable and legitimate. The VT Act was passed without significant opposition from the various economic powers.

The enactment of the VT produced an important change in the relationships between local governments and transport companies. Until the early 1970s, urban public transport had been in the hands of private operators under full risk agreements with the municipalities. When the VT was extended in the mid-1970s, the nature of these agreements gradually changed. Under the full risk contract, transport firms owned their assets (bus fleet, warehouses, workshops) and were paid in full from the farebox. This was no longer possible when operating deficits started to increase. Consequently, local governments bought operators' assets and entered into other forms of agreements with public transport companies who gradually became mere service providers (Lefèvre and Offner, 1993). Indeed, most of the contracts between public transport companies and authorities do not give any leeway to operators. Fares and fare systems are decided by the authority (but are monitored on a national basis by the government) and planning is also done by local governments. The company is, in theory, only responsible for the day-to-day provision of a service. Its remuneration is fixed in the contract, with deficits being paid by the authority from the VT. Competition between firms when contracts expire is rare and contracts are very often automatically renewed.

The system has not changed significantly since the 1970s, with only one minor development in the form of linking a small proportion of the operator's remuneration to management results (increase in passengers, stabilizing deficits, and so on). The public transport sector in France therefore has an original type of management, with services under the complete financial and political control of local governments in the provincial cities, and controlled by the state in the Paris area, but nevertheless operated by private companies for the most part (in 1993, with the exception of the Paris area, the 134 most important networks were operated as follows: 79 by private companies, 39 by mixed economy companies and 16 by municipal undertakings). As shown in Chapter 2, this situation is virtually unique in the developed world, but in France it is to be found in many other urban services such as water production and distribution, waste collection and treatment, cable, sewerage, and so on, to the extent that it has been called the French model of urban services management (Lorrain, 1991).

Soon after its inception, the VT was being looked upon as manna from Heaven by local governments which did not use it very wisely, funding projects and making up operating deficits which were no longer related to the journey-to-work. In 1986, the National Assembly extended the legal use of the VT to all public transport projects and operating deficits *per se*.

The 1971 Local Transport Tax Act established urban public transport as a specific, autonomous and privileged sector. While non-urban public transport declined steadily in the 1970s and 1980s, receiving no significant assistance from central or local governments, public transport in the urban areas flourished. As we have already seen, public transport supply has increased significantly in the last 20 years or thereabouts, and networks have been developed and modernized. This development was largely due to the VT since it was a dedicated tax and heavy enough to fund this development without local government or central government help. The VT allowed local governments to develop their networks while at the same time helping them maintain a systematic low fare policy, which means that the VT was heavily used to make up the increasing deficits.

Indeed, in the Paris area the portion of the VT used for covering operating deficits constantly increased in the 1980s (82 per cent in 1981 and 94 per cent in 1988) although this was slightly reduced to 89 per cent in 1992 (Observatoire Régional des Déplacements, 1993).

In the mid-1980s this happy period ended, since the productivity of the VT diminished as the necessity for greater investment in public transport infrastructure meant that it could no longer keep pace with the increase in deficits. Central government, transport authorities and operators agreed on an increase in the VT rates, but this proved insufficient. At the same time, following the example of other countries, a policy of productivity (decreasing costs and increasing revenues) and of diversifying sources of funding was introduced. From 1976 to 1986, the Ministry of Transport assisted public transport investment through development programmes (*contrat de développement*) which marginally served to buy buses (that is, to renew the fleets). They were converted into productivity programmes in 1986. These programmes were used to improve the productivity of the networks by increasing their commercial speed (through improvements to the street system) and their regularity (through the implementation of operating assistance systems), and also by making public transport more attractive to users (more comfortable buses, better passenger information, and so on). An increase in revenue was achieved mainly by raising fares, although public transport tariffs remained strictly controlled by the state. This policy proved successful during first years of its application, but now seems to have reached its limits, as the current lower growth rate or even stabilization of most productivity indicators show. On the investment side, it soon became crucial to find new sources of funding because of the VT shortage and the unwillingness (or inability) of local authorities to finance infrastructure from their own budgets. The participation of the private sector (through property taxes – value capture – or direct investment) seemed indicated (Quin *et al.*, 1990), but so far nothing has been achieved. In short, the creation of a more balanced transport system seems to have reached a standstill.

CONCLUSION

So far, the present situation of urban transport in France is the result of three major policy defects: (1) transport policies have been too sectoral and often conflicting; (2) crucial elements have been forgotten; and (3) they have lacked clear and comprehensive objectives.

Policies have been too sectoral and often conflicting

Multi-modal policies have never been the rule. Where central and local governments have launched and implemented policies, these have usually dealt with specific issues and have always related to a specific mode of transport, either the car or public transport. The essential comprehensiveness of the notion of travel has never been the focus of any concrete policy, despite much discussion. This lack of comprehensiveness is clearly illustrated by the political and managerial separation of responsibilities in the transport field. To put it bluntly, no organization or authority exists which may legally assume responsibility for a comprehensive transport policy. Municipalities and mayors are individually responsible for parking policies, traffic and street management, but public transport is administered overall by area-wide authorities and there is no connection between these two political powers. This situation, which is to be found in almost every country, has been largely condemned by the State and more recently by the Association of Public Transport Authorities, GART, which calls for a single area-wide authority to take charge of all transport modes and policy. The result is a set of sectoral actions or policies in every urban area which lack overall cohesiveness. In 1982, the LOTI established PDUs, one of their objectives being to make comprehensive transport planning a requirement for urban areas wishing to obtain government subsidies for transport. Every aspect and mode of urban transport was to be included in the PDU, the conception and implementation of which was the responsibility of the public transport authority which was required to show an integrated approach to transport issues. Very few urban areas followed the LOTI's recommendations and when, in 1986, the PDU was no longer required in order to obtain governmental assistance, the incentive to implement it disappeared. The lack of central, but also of local, political support for a long-awaited measure indicates that an important opportunity was unquestionably missed.

Not only have urban areas lacked comprehensive and integrated transport policies, they have also suffered from conflicting ones. There seems, for instance, to be an inherent contradiction between those policies which favour making city centres more accessible to cars (construction of off-street parking facilities, for example) and those intended to deal with city centre congestion. There is also a clear contradiction between those policies which favour the car (free-flow lanes, parking spaces offered by public or private employers, for example) and the expressed desire for public transport to increase its customer base. In the early 1980s, the term 'modal supplementarity' was a leitmotiv, meaning that transport policies should not assume that public transport and the car were in competition, but rather that they were complementary. Park-and-ride was the symbol of this approach but, so far, despite the needs and the opportunities, few facilities of this kind have been built. Local elected officials have proved unable to choose between public transport and the car, and existing policies reveal their indecisiveness.

The 'forgotten areas' of transport policies

At least three 'areas' have been forgotten by the policy-makers: the area-wide territory, the outer suburbs and alternative modes.

The transport tax (VT) has played an important role in the making of area-wide public transport authorities. However, it must be said that although French cities do possess public transport authorities whose responsibility extends beyond the core of the urbanized area, they do not generally encompass the whole of the relevant urban territory. Indeed, one of the objectives of the PDU was to study and plan the relevant territory – that is, to go well beyond the existing PTU – but once again this has only been achieved in a few agglomerations, notably Lorient and Grenoble (Lassave, 1987; Lefèvre and Offner, 1990). The PTU in these cities has been extended to include some peripheral municipalities, but in most urban areas this has proved impossible, very often for political and economic reasons. Moreover, area-wide authorities were also forgotten by the decentralization laws of 1982 and 1983. Since decentralization, responsibilities and powers have been devolved to municipalities, *départements* and regions, but not to the existing area-wide authorities which therefore continued to be dependent upon their member communes for their public transport policies, while (as mentioned above) these very same communes are the sole authority for local traffic and roads. Such was the situation, therefore, when in 1982 and 1983 a golden opportunity to create area-wide transport authorities (to deal with public transport, traffic, road maintenance, street use, parking, and so on) was missed, essentially because most local elected officials opposed it.

An increasing number of people and activities are concentrated in the outer suburbs of urban areas, and this trend is likely to continue into the foreseeable future. The vast majority of trips which begin or end in these suburbs are made by car. It would seem that public transport authorities completely forgot these areas, thinking that in view of their low density it was useless to develop specific public transport systems to serve them. This was obviously car territory. However, some experiments with flexible modes of public transport (minibuses, dial-a-ride systems, van-pooling, and so on) were initiated in the early 1980s, essentially because society at that time was very aware of energy consumption and dependence. The low cost-effectiveness of these experiments (although some succeeded in attracting significant usage) and their organizational complexity led to their gradual elimination at the end of the decade. Today, because they possess the know-how, public transport planners seem more interested in building light rail systems or busways between peripheral areas and urban cores than in working on innovative ways and policies to serve the peripheries, considering that the car is too well adapted to these areas. Ideologically speaking, a battle has been lost without having been fought, which is all the more damaging since these are the places where most of population will live in the near future and where public transport could play a social role since most of the 'socially disadvantaged' neighbourhoods are located in these areas.

This is not the only battle to have been lost, however. Today, two wheelers are no longer part of the French transport scene. This is confirmed by almost all recent household surveys. Like public transport in the outer suburbs, two wheelers have been forgotten, as if they were modes belonging to the past. Yet their revival was on the political agenda with the LOTI and the PDU. In the early 1980s, so-called alternative modes were to be developed. Urban areas such as Lorient, Annecy and Grenoble were keen to develop infrastructures but this proved insufficient to encourage their use. The same may apply to walking, which is definitely on the decline as a result of safety problems and the decentralization of people's homes and activities. No major policies have been directed at preserving and encouraging walking, despite the implementation of a few footpath networks in areas such as Chambéry or Angoulême.

A lack of clear objectives opens the door to the car

The inability of urban managers and politicians to set clear objectives for their transport policies is already evident from the preceding argument. The undoubted result of this lack of clarity is the increasing dominance of the car in French cities, which can be seen in almost every aspect of transport policy. Very often, technology has been used to avoid making a choice between public transport and the car. For instance, the development of Operating Assistance Systems for buses was preferred to the creation of busways in city centres, since the latter are politically more difficult to introduce because they take traffic lanes away from the car. The same is also true of automatic metros (VAL) which have been preferred (Toulouse, Bordeaux) to trams (Strasbourg) not only because they present a more modern image, but also, being underground, because they do not compete with the car for street use. The establishment of the transport tax meant that public transport policies were not dependent on central and local government budgets and, since local authorities did not have to bear the political and financial costs, it was possible to develop a pro-car and a pro-public transport policy simultaneously.

Today, things are less clear. First, the 1970 social consensus on the VT is falling apart, with companies calling for considerable reductions claiming that it reduces their competitiveness. Second, the VT is insufficient to cover the expected deficits and infrastructure costs (well over FF120 billion in the next decade) and a new source of revenue will therefore have to be found. In addition, the car's 'needs' are increasing at a tremendous rate and it seems to be receiving greater support than in the preceding decade. If this *laissez-faire* continues, it cannot but contribute to an increase in the already overwhelming dominance of the car.

5 The Netherlands: A Wise Country Overtaken by World Trends

With 438 people per sq. km, the Netherlands is one of the most densely populated countries in the world. Situated at the crossroads of Northern and Western Europe, it serves as gateway for a heavy flow of traffic and goods. Its geographical position is an important economic resource and access to transport nodes (seaports, airports, and large cities) is vital. However, the essential incompatibility of high population density and heavy traffic means that the Netherlands is currently facing dangerous levels of congestion, especially in the Randstad, the economic heart of the country. Although it is essential that people and goods be able to move easily across and within the country, the Dutch strongly resent the incursion of heavy traffic on their daily environment; thus environmental protection is a particularly sensitive issue.

If this is true at national level, it is even more so within urban areas. Dutch cities have long been famous for their 'taming' of the car and for being environmentally friendly, the result of long-term policies in favour of alternative modes (such as cycling). Nowhere else is the use of the bicycle higher. Bicycles are the second most important mode of transport in cities, coming just after the car and a long way ahead of public transport. Indeed, this is a specific feature of urban transport in the Netherlands.

The situation seems to be changing, however, as a result of a marked increase in car ownership and use in recent years. Congestion and air pollution levels have increased dramatically, so that by the late 1980s the government decided to introduce comprehensive policies to deal with these problems.

URBAN SPATIAL STRUCTURE

In 1993, the Netherlands had a population of 15.2 million. The population is unevenly distributed, the Randstad being the most urbanized area of the country with a population density of well over 1000 per sq. km. The high density of the Netherlands means that space is precious and must be used wisely, especially since about one-quarter of the territory is below sea level. It is a highly urbanized country with a town of at least 10 000 people within 10 km of any given point. The Randstad, a ring-shaped conglomeration of contiguous cities, encompasses about 6 million people, representing 40 per cent of the Dutch population. Almost all the major municipalities (Amsterdam, Rotterdam, The Hague and Utrecht) belong to the Randstad, and two of them (Amsterdam and Rotterdam) each have a population of about 1 million people. No single centre dominates the Randstad. It is a polycentric metropolis, and movements of people and goods are therefore multi-directional.

Suburbanization started in the 1960s and 1970s in spite of restrictive land use policies. As in many countries, large cities have experienced, to some degree, a loss of population. Data on commuting flows in the Amsterdam area reveals a 17 per cent increase in commuting trips between the suburban ring and the city centre from 1975 to 1985, and a 21 per cent increase in trips between the city centre and ex-urban area (Jansen, 1992).

TRENDS IN URBAN TRANSPORT

The last 1990 National Travel Survey (NTS) indicates that, on average, every Dutch person makes 3.7 trips per day, which represents a significant increase since 1980 (3.2 trips). In the most urbanized areas or towns, each resident made about 3.6 trips per day in 1992. Table 5.1. gives figures for these trips, categorized according to purpose. The evolution of mobility is following the same trends in the Netherlands as in most other countries,

Table 5.1 Percentage distribution of travel by trip purpose, national aggregate (as percentage of total travel), 1980–92

Purpose	*1980*	*1986*	*1992*
Work	27.4	23.9	23.4
Shopping	20.8	24.8	25.0
School	6.2	5.6	4.8
Leisure	31.1	32.1	32
Other	14.0	13.2	14.5

Source: Central Bureau of Statistics.

with a proportional decline in the number of work and school trips and an increase in shopping and recreational trips over the last decade or so.

As in all other industrialized countries, the car is the dominant mode of transport, although to a lesser extent than elsewhere. At national level, cycling is also still a very important mode of travel, both in urban areas and in the country as a whole. This is shown in Tables 5.2 and 5.3. Indeed, these tables reveal a significant contrast with other countries. Two major sets of observations can be made.

First, about one-quarter of trips are still made by bicycle, which means that this mode of transport is used more heavily in the Netherlands than in any other country in the industrialized world. However, the car still dominates since it is used for about half of all trips, which corresponds approximately to the average for industrialized nations. Consequently, public transport represents only a small (much less than 10 per cent) percentage of trips except in the largest urban areas (Amsterdam, Utrecht).

Second, with regard to short trips of less than 20 km, which account for most urban trips, only two modes, the car and the bicycle, have shown an increase in the last decade. Although the use of the car for journeys of less than 7.5 km remains more or less unchanged, it is increasing dramatically for longer trips. The car is dominant for all trips over 2.5 km. Bicycle use

Table 5.2 Modal split in four Dutch urban areas (1986–90 travel surveys) and in other urbanized areas (National Travel Survey, 1992) (% of all trips)

	Amsterdam 713 000	*Utrecht 233 000*	*Breda 127 000*	*Zwolle 97 000*	*Urbanized rural municipalities**	*Towns*	*Total nation*
Bicycle	20.7	28.4	25.5	34.1	26.7	26.7	27.8
Walking	26.2	22.1	17.8	15.3	13.9	18.7	16.4
Car	36.0	38.3	50.7	45.8	52.8	43.7	47.7
Public transport	15.0	9.1	5.1	2.8	4.5	7.5	5.9
Other	2.0	2.0	1.2	1.8	2.3	1.4	2.0

*Urbanized rural municipalities: municipalities with an agricultural working population of 20% or less and specific commuter districts with more than 30% of non-resident commuters.

Source: Elaborated by the authors from CBS in Ministry of Transport, *Facts about Cycling in the Netherlands* (1993).

Table 5.3 Evolution of modal split according to the average trip distance, 1980–90 (%).

Trip distance: 0–1 km	*1980*	*1990*
Bicycle	29.2	32.1
Walking	59.2	58.9
Car	10.5	8.9
Public transport	0.1	0.0

Trip distance: 1–2.5 km	*1980*	*1990*
Bicycle	43.5	45.8
Walking	21.9	20.8
Car	31.6	30.2
Public transport	0.9	1.0

Trip distance: 2.5–5 km	*1980*	*1990*
Bicycle	33.9	36.2
Walking	7.5	6.9
Car	50.3	51.7
Public transport	4.5	3.5

Trip distance: 5–7.5 km	*1980*	*1990*
Bicycle	22.8	24.4
Walking	4.1	4.9
Car	63.1	63.4
Public transport	6.1	4.9

Trip distance:7.5–10 km	*1980*	*1990*
Bicycle	16.8	15.4
Walking	2.2	0.0
Car	64.1	76.9
Public transport	11.8	7.7

Table 5.3 Continued

Trip distance: 10–15 km	*1980*	*1990*
Bicycle	12.1	11.1
Walking	1.3	0.0
Car	71.8	77.8
Public transport	10.5	7.4

Trip distance: 15–20 km	*1980*	*1990*
Bicycle	6.9	7.1
Walking	0.4	0.0
Car	75.7	78.6
Public transport	12.6	14.3

Note: Data relates to all trips made by people over 12 years old and not during holiday periods.
Source: Central Bureau of Statistics.

is growing for all trips under 7.5 km and it is the dominant mode for trips of 1–2.5 km. It remains the second most important mode of travel for all other trips, except those of 15–20 km. Finally, public transport has declined significantly for almost all categories of trips, with the exception of distances of 15–20 km.

A major conclusion may be drawn from these remarks. The importance of cycling as a mode of urban travel is not affecting car use, which continues to grow. Public transport use, however, is declining because passengers are being taken away by the bicycle and the car.

Car ownership and use in the Netherlands have increased dramatically over the last two decades. From 1970 to 1992, the number of cars per 1000 people almost doubled, rising from 195 to 374. Car use has followed the same trend with 15.9 billion passenger-km of traffic in 1960, 66.3 billion in 1970 and 128.1 billion by 1992. Moreover, the Second Transport Structure Plan (Second Chamber of the States-general, 1990) forecasts a saturation rate of car ownership by 2010 with a 70 per cent increase if the trend experienced in the last decade continues. To accommodate this increase the road network has been expanded, with most of the work taking place in the 1960s and 1970s. Ring roads were built around

Amsterdam and Rotterdam. Although the motorway network has been strengthened in recent years (there was a 28 per cent increase in road length between 1978 and 1992) and had reached 2118 km by 1992, no major projects are now on the agenda.

Cycling is the second most important urban transport mode in the Netherlands and for this reason deserves specific study. Bicycle ownership has increased steadily over the last 40 years. While about 6 million people owned a bicycle in 1950 and 7 million in 1970, today their number has reached 12 million, which means that four out of five Dutch people are bicycle owners. However, bicycle use has not evolved parallel with bicycle ownership. In 1960, no fewer than 17.1 billion passenger-km were travelled by bicycle. This number gradually fell until the mid-1980s (about 12 billion passenger-km in 1986), but has recently increased slightly, reaching 12.8 billion in 1990. There are many explanations for the renewal of bicycle use, one being the tremendous development in the building of bicycle infrastructure. From 1978 to 1992, the length of cycle lanes and paths almost doubled, rising from 9282 to 18 175 km, extending the use of the bicycle to cover every type of trip. While in most countries the bicycle is used for recreational activities, in the Netherlands it represents 27 per cent of commuter trips and one-third of shopping trips (Ministry of Transport, 1993).

The success of the automobile and the relative success of the bicycle have been detrimental to public transport. Whilst it is true that public transport in the 49 major urban areas has shown a slight increase in use in terms of the number of trips and passenger-km since about 1985 (Table 5.4), this growth is significant in volume only, not in the share of total urban trips. However, the weakness of public transport is not due to the lack of public transport infrastructure in urban areas.

Table 5.4 Urban transport usage changes, 49 major cities, 1981–92 (in millions)

	1981	*1985*	*1990*	*1991*	*1992*
9 largest cities					
trips	578	533	584	631	635
passenger-km	2 209	2 085	2 069	2 259	2 287
40 other cities					
trips	87	83	116	142	147
passenger-km	313	299	477	577	635

Sources: Koninklijk Nederlands Vervoer, Kerncijfers Personenvervoer, corresponding years.

While most Dutch cities had tram networks before the Second World War, only three (Amsterdam, Rotterdam and The Hague) still had them in the 1970s. From 1970 to 1980, these tramway networks have been upgraded and extended and metro lines have been added to existing networks. The first underground train line in the Netherlands opened in Rotterdam in 1968. It was followed in 1977 by the first line of the Amsterdam metro, and in 1979 by a new LRT line in Utrecht. Since then, one additional metro line has been opened in Rotterdam, and one combined LRT/metro line in Amsterdam.

URBAN TRANSPORT PROBLEMS

Air pollution and road congestion are, without a doubt, the two most important problems Dutch cities will have to face in the coming decade. This is clearly stated in the Second Transport Structure Plan, which emphasizes environmental deterioration and the decreasing level of accessibility of urban areas, especially the Randstad. This does not mean that other problems common to most cities are not to be found in the Netherlands (for instance, public transport financing), but they seem either to be less significant or to have already been dealt with successfully (road safety, for instance).

Air pollution has continued to worsen in the last decade, mostly because of the tremendous increase in car traffic. Indeed, 92 per cent of NOx emissions are due to passenger travel, and 91 per cent of total carbon dioxide emissions are produced by cars. The success of policies to stabilize NOx emissions by 1980 now seems to be offset by traffic growth.

If nothing is done to restrain car ownership and automobile use, the cost of congestion (estimated at about 1 million Dfl in 1990) will increase fourfold by 2010 and the Randstad will become increasingly gridlocked. It is a question not only of car traffic, but also of freight traffic. Congestion is not an urban problem *per se*, since the inner cities are not significantly congested, but rather an interurban problem. Considering the density of the country and the short distances between the major urban centres, interurban congestion would be considered 'urban' by international standards. Most congested sections of the road network are on the fringe of metropolitan areas. About 72 per cent of delays are caused by the inadequate road capacity (Korver *et al.*, 1992).

As in most other countries, the problem of road congestion will not be solved simply by increasing road capacity, and a shift towards public transport will be necessary. However, public transport networks are not sufficiently developed to be able to carry a much larger volume of passengers. An already saturated national railway network has become even more overcrowded since the introduction of a national student pass in 1991, and new rail infrastructure and facilities will have to be built in the Randstad to allow the national railways (Nederlandse Spooswegen, or NS) to carry the increasing number of passengers. While only 205 million passengers were carried in 1981, usage increased to 256 million in 1990 and had risen to 333 million by 1992: a 30 per cent increase in the last three years.

On the other hand, Dutch roads are among the safest in Europe. While the Netherlands registered 3181 traffic fatalities in 1970, by 1992 this figure had been reduced to 1285, about the same rate per 100 000 inhabitants as the UK.

Taking the country as a whole, public transport does not represent a large proportion of urban trips. However, in some areas (mostly the largest cities), it does represent a substantial proportion of travel (Table 5.3). Transport companies have been publicly owned for a very long time, with some municipal operators (notably in Amsterdam) having been bought by the municipalities in the early 1900s. The organization of public transport is as follows. The nine largest cities (Amsterdam, Arnhem, The Hague, Dordrecht, Groningen, Maastricht, Nijmegen, Rotterdam and Utrecht) own their municipal operators. Forty 'secondary' cities are served by regional companies, most of which are owned and controlled by the central government under specific contracts with the municipalities. These regional operators also serve rural areas. All public transport companies have been subsidized for more than two decades. The public transport deficits of the 49 urban networks have been only partially reduced over the last decade (Table 5.5) and even this has proved to be a fragile success, with recent figures showing a stabilization of deficits.

TRANSPORT POLICIES

The Dutch define their political system as 'a unitary decentralized system'. However, in the field of transport, centralization seems to have been a major feature of policies in the last 30 years or so, even though localities have participated in the policy-making process. Centralization is therefore the general framework within which most policies have been decided and implemented. In the first part of this section we will describe the context of political and institutional policy-making before going on to examine the major sectoral policies introduced over the last decade, especially as

Table 5.5 Public transport deficit, 1981–92 (in current million Dfl and in % of total cost)

		1981	*1985*	*1987*	*1990*	*1991*	*1992*
9 largest cities	M Dfl	902	1 035	1 064	1 090	1 083	1 075
	%	79	77	78	77	73	71
40 other cities	M Dfl	116	150	148	207	225	239
	%	75	76	72	74	72	72
Rail	M Dfl	818	1 287	1 345	1 394	1 431	1 472
	%	47	54	54	52	50	44
Total public	M Dfl	2 437	3 121	3 177	3 377	3 479	3 534
	%	63	63	63	61	54	55

Sources: Koninklijk Nederlands Vervoer, Kerncijfers Personenvervoer, corresponding years.

regards the environment (air pollution and spatial environment) and traffic conditions.

In the Netherlands most of the financing, as well as the power system, is concentrated in the hands of central government. This is true not only in the field of transport but in most other domains as well. For instance, most local government revenues come from central government through categorical grants and a block grant (called the Municipality Fund). In the transport sector, centralisation indicates that central government is by far the largest financing organization and that it controls most public transport companies and carries most of the responsibility.

Until the early 1970s, municipalities played an important role in public transport. They controlled the operators, set the fares and financed the deficits. The Ministry of Transport had to subsidize regional operators and the national railways. Because of the tremendous increase in public transport operating deficits in the 1970s, central government had to intervene and replace the municipalities which could no longer pay them. This was done by withdrawing public transport financing from the Municipality Fund. Since the level of deficits was partly the result of fare policies and since these policies were decided by the municipalities, the tension between the state and local governments grew. Those who decided fare policies were no longer responsible for their financial impact. Thus, in the mid-1970s, the Ministry of Transport took over responsibility for setting fares and decided to implement a new nation-wide fare structure. This was imposed in 1979 by introducing fare zones all over the country, which required the adjustment of local fare systems since some municipalities had fares which were higher or lower than the proposed national fares. After regulating fares, the Ministry of Transport decided to control the operating costs of the various public transport companies. It established national standards from which actual deficit compensation was set and imposed austerity measures (fare increases, staff reduction, and so on). In so doing, it increased its control over the whole public transport sector but so far, operating deficits have been only partially reduced.

One major drawback to the existing organizational and financing system is the operators' lack of responsibility. The Ministry of Transport considers that such a system does not encourage public transport companies to search for more passengers and that a change in the rules of the game is therefore necessary in order to introduce some competition into the sector. Until now, public transport companies – those owned by the municipalities as well as those controlled by the state, such as the regional operators – have had a 'territorial' monopoly in serving their specific areas, and this system is judged counterproductive. It means, for instance, that in a city, only the municipal company (if it exists) or the regional company which has a contract with the local authority can operate, and local authorities cannot sign a contract with other companies. To solve this problem, certain deregulation measures are now being seriously considered. The introduction of a system of 'comprehensive' competitive tendering for each transport area is envisaged, for example. Under this system, all public transport companies as well as private operators will have to compete for a whole service network (routes and levels of service will not be specified by the transport authority but, on the contrary, will be established by the bidders). There will be no more territorial monopolies. Unlike the British system, however, competition will not be *on* the road but *for* the road. A first experiment will be launched on interurban services in 1995 and, if successful, should be implemented on urban services in the near future.

So far, centralization has not proved successful in reducing the public transport deficit or in significantly increasing ridership. The situation of public transport supply is no better. As previously noted, a fixed public transport infrastructure is most certainly lacking in some cities, especially in the Randstad. The difficulties of the Amsterdam metro, which has proved five times more expensive than forecast and which has a reputation for vandalism, added to the public finance crisis and did not help in launching a policy of investment. In 1990, recognizing the need for further investment to finance the newly established role of public transport in improving the accessibility of the Randstad, central government authorized a substantial funding programme of 20 billion Dfl over 15 years.

Many cities have already prepared various projects for metro systems or LRT (known as 'Snelltram', to avoid the negative connotations of the Amsterdam metro). So far, besides the North/South Snelltram line in Amsterdam, no significant project has been approved, mostly because of the state's austerity policy. The situation is no better for the NS railway network, part of which serves urban areas and which, because of the smallness of the country, its high density and level of suburbanization, is being used increasingly for commuter trips. The establishment of a virtually free national student pass in 1991 has put additional pressure on the network by increasing patronage and deficit at the same time. The NS has launched a national plan, Rail 21, to augment the capacity of the rail network, notably in the Randstad where congestion problems are most serious.

The decentralization of public transport policies has been on the government agenda for almost a decade. A first measure taken at the end of the 1980s was intended to give municipalities some flexibility in deciding on the fares that should be adopted in their area. The Ministry of Transport established a price range, and municipalities were allowed to set fares within this range. With deregulation under way, fares will be proposed by the bidders, but these too will have to be set within a framework established by the government. New decentralization measures are to be introduced shortly with the establishment of transport regions. Transport regions are areas where a single transport authority, similar to the French model of *autorités organisatrices* (organizing authorities), will be established. Responsibility will then be transferred from the municipalities and central government to these public transport authorities. Such a project is already well advanced in the Amsterdam and Rotterdam areas.

The sharp increase in car ownership and use has created tremendous problems with regard to air pollution and traffic conditions. Finding a solution to both these problems has been on the government agenda for several decades now, and various measures to reduce air pollution caused by the automobile have been enacted in the last decade. In 1989, for instance, 71 per cent of new cars sold in the Netherlands were equipped with a three-way catalytic converter, and tax incentives for the cleanest vehicles were introduced so that new cars are now no more expensive than non-equipped ones. These measures have proved successful. NOx emissions, for example, which had increased up until 1980, have since remained stable. Total emissions of CO increased until 1970, after which they stabilized and have now decreased to about the 1960 level (Ministry of Transport, 1993). However, these encouraging results are now being jeopardized by the tremendous growth in automobile traffic. Carbon dioxide emission is still rising, with the result that air pollution in the Netherlands is increasingly being considered as a crucial issue which cannot be solved by technological measures alone.

Road congestion is also an important issue, notably in the Randstad area. To prevent a total gridlock of this part of the Netherlands, a Plan of Accessibility has been accepted. Having decided that, in view of the social and political opposition, building new motorways was neither a satisfactory nor a feasible solution, the government decided to rely on technology and public transport to make the Randstad more accessible. In order to increase the capacity of the road network, experimental traffic management schemes have been implemented in Amsterdam, Utrecht and Breda; evidently with some success, since the average commercial speed on the networks has increased by 5 per cent. To reduce traffic, electronic road-pricing was envisaged in the 1980s but the law was rejected by the National Assembly. Today, the introduction of a tax to use the Randstad network is envisaged for 1995. Other drastic measures to reduce automobile traffic are now being planned. These include restrictive parking policies (reduction of the number of parking places in city centres and an increase in parking fees) or, in the longer term, the reduction of car mobility by the spatial concentration of residential and work places and the development of teleworking.

Dutch urban areas do have serious transport problems. However, they are also well known for being among the most pleasant urban places to live in in the world. Urban sites have been well preserved, and walking and cycling are agreeable modes of transport. This quality of life may, in part, be the result of the Calvinistic element in Dutch society, but it is clearly also the result of comprehensive policies implemented at local level.

Dutch cities are famous the world over for the concept of *woonerf*, introduced in the 1960s and early 1970s. *Woonerven* are residential streets or areas reconstructed as 'shared surfaces' for the various transport modes. The rationale for these experiments was to 'civilize' traffic in residential areas. To do so, the distinction between footway and carriageway was abolished and streets were redesigned to reduce the speed of cars, usually to a maximum of 20 km/hour. These experiments have gradually been extended to shopping areas and village centres. Furthermore, many Dutch towns (Groningen, Zwolle and Enschede were pioneer cities in this respect) have implemented restricted vehicle areas and car-free zones, first in historic towns such as Delft, Haarlem and Leiden and then in larger areas. Traffic-calming areas are therefore common in most Dutch cities.

All these measures have fostered the development of the use of alternative modes like the bicycle. In order to maintain the level of bicycle use, cycle lanes and paths have been built, but their success is due not only to the large-scale development of infrastructure but also to the way they have been designed, most of them being integrated into a more comprehensive policy. The bicycle, for example, has very often been combined with other modes such as the train, with bicycle storage facilities in railway stations and the establishment of integral bicycle networks which make it possible to cycle from one place to another within urban areas without being troubled by car traffic. In this respect, Dutch policies on bicycling should be carefully examined by countries such as France and the UK, where the use of the bicycle has drastically declined over the last decade.

Consequently, it is hardly surprising that Dutch roads are among the safest in the world. However, Dutch people and the national government consider that improvements are still necessary and possible: for instance, 30 km/hour zones were established in many urban areas in the mid-1980s and they have evidently proved successful. A survey done in 15 municipalities to assess the effects of such measures showed they had brought about a 15 per cent reduction in the number of accidents and a diminution in the number of road deaths, notably among cyclists. Although declining, the volume of road fatalities is still considered too high. The government has set a reduction of 25 per cent as the objective to be reached by 2000.

CONCLUSION

The importance of the bicycle in the modal split and the successful 'taming' of the car in many urban centres are undoubtedly achievements for Dutch cities. However, these achievements are still fragile. The automobile has become the most important mode of transport in urban areas and has not suffered significantly from the development of the bicycle, which seems to show that there are other rationales for car use. The zero sum game is not being played between the car and public transport but between the bicycle and public transport. The fact that it is difficult to increase the use of one without decreasing the use of the other seems to be confirmed by looking at the evolution of modal split in other countries where public transport use has been increased to the detriment of alternative modes. Public transport does not represent a high proportion of urban trips, and although this can also be considered characteristic of the Netherlands, it leaves politicians and technical experts with a 'ridership reserve'. Indeed, one of the most important challenges in the near future will be to develop public transport supply and use without reducing bicycle use. This cannot be done without harming the automobile. Will the Netherlands take the lead in this respect? If local authorities or central government do not act, the Netherlands may soon find itself on the same road to dependence on the car as most other countries.

6 Italy: Cities without Policies

The urban transport crisis in Italy is very serious and generally considered to be an acute problem in the functioning of cities. The high degree of pollution and congestion are characteristic of the crisis but, to a far greater extent than in other countries, it seems to be a crisis of public policies or rather, absence of policies: an absence of policies designed to achieve a balance between public and private transport, and an absence of policies in the public transport sector itself. The lack of infrastructure, notably in public transport, is criticized by many. The inability of successive governments to reduce the increasing deficits incurred by public transport operators is increasingly evident today. Public bodies, from the state down to local governments, do not seem to have any policies (that is, any objectives at all), let alone any coherent policies. This is corroborated by the fact that no recent comprehensive data on urban transport is to be found. Public action is generally determined by the degree of urgency of a problem, but so far it has been sporadic, sectoral and short sighted. This situation is undeniably the result of fragmented political power and political instability over at least the last decade, which seems to indicate that it is politically less risky to do nothing than to do something. This analysis will be developed later, but we will first try to present a general picture of the state of urban transport from the scanty and incomplete information which does exist.

TRENDS IN URBAN SPATIAL STRUCTURE

Urban areas in Italy have experienced a slight increase in population growth over the last two decades (see Table 6.1). Generally speaking, this increase has taken place largely in the outer suburbs, but a distinction must be made between Northern and Southern Italy. Between 1971 and 1981, suburban growth was greater in the South than in the North and the decline of city centres was greater in the North than in the South, with some urban areas like Palermo, Rome or Bari showing a substantial increase. Since 1981, the difference between Northern and Southern cities has widened, but against a background of slower population growth.

While, on average, Southern urban areas have maintained a slight increase in population, Northern areas have shown a decrease. This decrease has taken place essentially in city centres and, even in the South, the population in city centres has begun to decrease slightly.

Urban areas have therefore expanded over the last two decades at a different pace in the North and in the South. However, despite this evolution, city centres have remained the greatest pole of attraction in urban areas.

TRENDS IN URBAN TRAVEL BEHAVIOUR

There is little recent data on urban mobility in Italy. Most of the existing information is derived from the already outdated 1981 population census, the 1991 data on urban travel not being available yet. Since 1981, very few surveys have been conducted on this topic and most of them, like the 1986 Censis survey or the 1990 Iveco study, are incomplete and only partial in scope. The 1986 Censis survey, for instance, only deals with four urban areas (Bari, Bologna, Rome and Milan) and is limited to motorized transport. The 1990 Iveco study also covers only four areas (Genoa, Milan, Naples and Turin) and excludes walking and cycling. The annual

Table 6.1 Population change in the largest Italian metropolitan areas, 1971–1987

		1971–81 change (%)			*1981–87 change (%)*		
	pop 89 millions	*Central city*	*Suburban areas*	*Whole area*	*Central city*	*Suburban areas*	*Whole area*
Turin	1	–4.35	+14.92	+2.24	–8.21	+4.43	–3.35
Milan	1.4	–7.35	+13.74	+2.33	–7.87	+4.07	–1.78
Genoa	0.7	–6.61	+10.52	–5.31	–5.36	–0.4	–4.92
Bologna	0.4	–6.41	+21.25	+0.51	–6.94	+5.89	–3.06
Florence	0.4	–2.07	+14.51	+5.07	–6.03	+3.36	–1.62
Rome	2.8	–2.09	+32.70	+4.63	–0.81	+10.47	–0.37
Naples	1.2	–1.16	+22.94	+9.13	–0.94	+9.7	+4.17
Bari	0.4	+3.85	+22.14	+9.84	–3.27	+11.5	+2.11
Catania	0.4	–4.93	+48.51	+8.78	–2.13	+20.18	+5.68
Palermo	0.7	+9.17	+15.18	+10.60	–3.86	+12.13	+5.14

Sources: Censis and Istat.

data compiled by the Ministry of Transport in the Conto Nazionale dei Trasporti (CNT) is restricted to public transport and private cars.

What information can be gleaned from these incomplete studies? First, the mobility of the average urban resident seems to have remained constant over the years at 3.6 trips per day. Second, mobility is described as erratic, meaning that people's movements in time and space are unpredictable, which seems to favour the most flexible means of conveyance: the car. The *macchina* is more and more heavily used by urban Italians as the various CNT studies (Table 6.2) show.

Erratic mobility is reflected in the decreasing number of trips related to work or schooling. Today, these trips account for only about one-third of the mobility of the urban population as confirmed by the Iveco study (1990). Data on modal split is also scarce and difficult to compare. The CNT study published in 1984 from the 1981 census data covered 13 metropolitan areas only, and dealt solely with work- and school-related travel. These trips were made principally on foot or by two wheeler (54 per cent), the car representing about 25 per cent, and public transport (bus and rail) 21 per cent, of the total. The 1986 Censis survey on motorized transport to work or school showed a very different picture, with the car accounting for most of the trips (about 61 per cent), bus and rail 27 per cent, and walking and two wheelers only 12 per cent. As with the Iveco study (1990), it presents the modal split as shown in Table 6.3.

General trends in urban mobility in Italian cities are therefore not easy to describe. However, all existing studies or surveys seem to indicate clearly that public transport is declining in favour of the car, which is understandable in view of the way in which urban trip patterns are evolving (erratic mobility). In this respect, Italy seems to be following the general trend of OECD countries described in Chapter 2.

The growing use of the automobile in urban areas is obviously related to the increase in car ownership over the last two decades (Table 6.4). This

Table 6.2 Evolution of modal split in urban areas, 1984–90 (% of passenger km)

	1984	*1986*	*1988*	*1990*
Public transport	14.3	13.2	11.2	9
Car	85.7	86.8	88.8	91

Source: Ministero dei Trasporti; Conto Nazionale dei Trasporti various years.

Table 6.3 Modal split in four urban areas in 1990 (% of motorized trips)

	Public transport	*Car*
Genoa	41	59
Milan	45	55
Naples	35	65
Turin	45	55

Source: Iveco (1990).

Table 6.4 Population (in 1000) and number of cars (in 1000), 1970–1992

Year	*National population*	*National registration*	*Cars per 1000 inhabitants*
1970	53 832	10 209	189
1975	55 589	15 061	270
1980	56 479	17 686	313
1985	57 202	22 495	393
1990	57 746	27 416	474
1992	58 163	28 500	490

Source: Elaboration by the authors from Ministero dei Trasporti, CNT, various years, and Eurostat (1993).

has increased about three-fold since 1970 with figures for the 1980s showing a notable discrepancy between the North and the South of the country, the North and Centre having about one car per two inhabitants and the South roughly one car per three inhabitants. Figures for the 1990s show an average of about 490 passenger cars per 1000 inhabitants, which places Italy at the forefront of European countries in terms of car ownership. Not only is car ownership high, but the percentage of households having two or more cars is also one of the highest in Europe and stood at about 20 per cent in 1987–88 (Salomon, Bovy and Orfeuil, 1993). However, the percentage of households not owning a car is still high, representing about one-quarter in the same year.

The only available data for urban areas is to be found in the 1986 Censis survey which does not seem to corroborate national data for the North–South split, since Northern cities do not have significantly higher car ownership rates than Southern cities (Table 6.5). Data on automobile use at the urban level is also very scarce. The various national transport surveys carried out in the CNT show an increase not only in the percentage of trips made by car in urban areas, but also in absolute figures (about 102 billion passenger-km in 1983 and about 164 billion in 1990), which points to a reverse trend from that indicated by the national data (72.1 per cent of medium and long-distance trips in 1985, but only 70.6 per cent in 1992: Ministero dei Trasporti, CNT, 1993). More specific data is very difficult to obtain and this information cannot represent the real urban traffic level of the automobile since it includes only traffic on provincial and municipal roads (part of it may not be urban at all) and excludes urban traffic on national roads and motorways.

As shown in previous tables, the role of public transport has declined significantly over the last decade. This is true in absolute terms, but it varies significantly according to the different modes of public transport. In Italy, local public transport encompasses two types of operator and four types of transport: (1) the 28 local rail companies either owned by local government or controlled by private firms (the *Ferrovie in concessione ed in gestione governativa*); (2) the 1139 companies (*aziende*) which operate (i) local bus services in urban as well as 'extra-urban' areas, (ii) streetcars (in urban and extra-urban areas) and underground systems. In urban areas, these companies are generally owned by the municipalities, while in extra-urban areas most of them are privately owned. Data on public transport is expressed in terms of technical and institutional modes such as municipal buses, tramways, underground systems, rail networks controlled by provincial and regional governments and the Ferrovie dello Stato (FS), the state rail company. No comprehensive public transport figures are therefore to be found. However, a synthesis, albeit incomplete, has been

Table 6.5 Car ownership, 1986 (% of population having at least one car in metropolitan areas)

City	%
Turin	55.5
Milan	58.8
Venice	38.5
Genoa	47.6
Bologna	58.8
Rome	58.8
Naples	40.0
Palermo	47.6

Source: Censis (1986).

Table 6.6 Urban public transport usage, 1981–91 (in passenger millions and in %)

	1981	*1985*	*1991*	*1981–91 (%)*	*Modal split (%) in passenger/*	*passenger-km*
Bus	4 136	4 009	2 905	–28*	77	75
Tram	493**	439	402	–18	11	10
Underground	310**	424	504	+62	11	16
FS	14***	11	9	–34		< 1

*1985–91.
**1980.
***1983.
Source: Table made by the authors from Conto Nazionale dei Trasporti, various years and Ministero dei Trasporti (1993a).

attempted in Table 6.6. It is incomplete first because it does not include the sector of local rail networks not controlled by the state (FS). This sector mainly provides an intercity service but, in some cases, as in Milan, Turin and Naples, it plays a crucial role, notably in commuter travel. Although no comprehensive data regarding the importance of these networks in urban travel exists, their presence must be acknowledged. Although the Naples (Circumvesuviana) and Turin (SATTI) networks did show considerable ridership declines (–25 per cent and –13 per cent respectively) between 1985 and 1991, the Ferrovie Nord Milano (FNM) experienced a substantial growth in use of 15 per cent thanks to an active policy on the part of the Lombardy region. The table is also incomplete because, although the decline in public transport use has been considerable (–23 per cent) in the last decade, this situation applies mainly to bus networks since the other declining forms of transport, tramways and the FS represent only a small part (tramways) or a very marginal part (FS) of public transport use. The dramatic increase in use of underground networks resulting from the opening of new lines during this period has been largely insufficient to offset the general decline of other means of transport.

THE URBAN TRANSPORT CRISIS

Urban mobility is in deep crisis in Italian cities. This is clearly stated by all sources, either private (Fiat, Iveco) or public (Ministry of Transport, local governments), and is true of all forms of transport. Italian cities are becoming more and more congested and more and more polluted, and safety is also becoming a major problem. Public transport is also in very bad shape with fewer passengers and ever-increasing deficits. This situation is not new, but the various measures projected and sometimes undertaken (see below) have not proved successful.

Congestion and pollution problems are the current leitmotiv of urban transport, but so far no clear information concerning them exists. National, or even local, data concerning metropolitan areas is nowhere to be found. The only existing data refers to municipal areas which is unsatisfactory because congestion and pollution are not simply municipal problems, but are, at the very least, metropolitan (area-wide) issues. Pollution came on to the political agenda when alerts were launched in many Italian cities, as in January 1993 in 11 urban areas. As for the increase in congestion, this is obviously the result of the dramatic growth in car traffic with almost no new infrastructure having been built over the last decades. This situation has also brought about a rise in the number of urban fatalities. Out of 100 accidents, 73 occur in urban areas (Istat and Automobile Club d' Italia, 1992). No fewer than 42 per cent of deaths and 70 per cent of injuries occur in cities. In 1991, there were 166 000 urban injuries and 3189 urban fatalities. This number has increased considerably since 1988 when there were 2825 deaths, while at national level the number of fatalities has constantly declined.

Urban public transport is also in a shambles. Investment in this field has been almost non-existent over the last two decades. Lack of money has meant that road transport companies have been unable to renew their fleets, so today, the average age of the fleet is constantly rising (Iveco, 1993). There are few underground systems and these have developed extremely slowly. As we have shown, Italy has several urban areas with a population of over one million. To date, however, underground systems exist in only two major cities: Milan and Rome. In Milan, the underground network is just 70 km long with only 3 lines, the third of which was opened in 1990. In Rome, the undergound network comprises two lines and covers 33 km. Many extensions have been proposed but so far none have been built, with the exception of a 5 km extension now under construction. An underground line does exist in Naples, but it is only 4.5 km long, for a city of over 2 million people. An underground system for Bologna is in the planning stage, but this is unlikely to open before 2010. Tramways (*tranvie*) were eliminated in most Italian cities except Milan and Turin in the 1950s and 1960s but, with the very recent exception of Genoa, no new LRT lines have been implemented. Regional railways (the *ferrovie in concessione* as well as the FS commuter lines) are also very run down, their lines congested from the necessary mix of passenger and freight services, which results in a very low commercial speed and, consequently, unreliability. This absence of investment is clearly the result of a shortage of public money, the bulk of which comes from central government which has proved unable to carry out a programme of investment in local railways despite the many legislative measures (see below) taken since 1980.

Consequently, it is not surprising that ridership has declined since many lines are badly congested, and public transport is less and less able to provide residents with an adequate service. The physical degradation of public transport networks (with some exceptions, such as Milan) has been accompanied by a substantial growth in operating deficits (see Tables 6.7 and 6.8).

In 1993, the total accumulated local public transport deficit amounted to about 12 000 billion lire ($8 billion), this being the deficit remaining after government operating subsidies. To be more specific, it is the amount of accumulated deficit since 1987 not covered by central government subsidies. In fact, since 1985 (Ministero dei Trasporti, 1993b), revenue from fares has covered about one-fifth (22.5 per cent) of operating costs, and subsidies 57.2 per cent, the remaining 20.3 per cent being covered by the operators themselves by borrowing in the financial market. At present, the remaining deficit is increasing by about 2000 billion lire (about $1.5 billion) a year. The bulk of the local public transport deficit comes from the major urban areas: about 70 per cent of this deficit is produced in

Table 6.7 Fare recovery ratio of various local public transport modes, 1985–91 (%)

Year	*Rail**	*Buses***	*Trams*	*Underground*
1985	13	26	23	31
1987	13	26	29	38
1990	12	23	26	30
1991	13	22	30	29

*Local networks of the 28 private or semi-public companies (*Ferrovie in concessione*). Most services are local but non-urban.
**Includes urban, as well as non-urban services.
Source: Conto Nazionale Trasporti in various years and Ministero dei Trasporti, (1993a).

Table 6.8 Operating deficits of urban bus networks by region, 1990 (in million lire and in %)

Regions	*Expenditure*	*Revenue*	*Ratio*
Piedmont	244 223	90 597	37
Lombardy	499 246	148 925	30
Trento	17574	4840	27
Bolzano	16 614	4075	24
Veneto	218 807	65 034	30
Friuli	97 578	33 580	34
Liguria	329 045	105 610	32
Emilia-Romagna	280 213	89 465	32
Total North	1 703 300	542 126	32
Tuscany	291 210	71 312	24
Marches	58 778	20 558	35
Umbria	45 960	12 009	26
Latium	1 119 610	149 798	13
Total Centre	1 515 558	253 677	17
Abruzzo	46 161	13 840	30
Molise	7 282	1 851	25
Campania	676 561	68 773	10
Apulia	170 791	26 596	16
Basilicata	24 150	3 214	13
Calabria	44 795	9 891	22
Sicily	398 227	69 978	18
Sardinia	93 449	21 538	23
Total South	1 461 416	215 681	15
National total	4 680 274	1 011 484	22

Source: Table adapted by the authors from Ministero dei Trasporti (1993a).

the largest urban areas with some important geographic variations, as shown in Table 6.8 with regard to local bus companies.

It is obvious from Table 6.8 that Northern regions show higher fare recovery ratios than Central or Southern regions. Some studies (Ministero dei Trasporti, 1986) indicate that costs, notably employee costs, vary significantly from one region to another, Southern areas showing more employees per bus-kilometre than Northern parts of the country.

THE POLITICAL INABILITY TO IMPLEMENT POLICIES

Urban transport is in a shambles, but the situation is not new. According to the Fiat Report of 1989:

> Many years ago, experts and operators in the sector called attention to the dangerous state into which urban transport would fall if comprehensive and urgent measures were not taken. In fact, this has had no effect. The situation has been one of constant deterioration and today, the problems can no longer be dealt with by superficial means. Lately, urgency has replaced unconcern, with improvised and often contradictory measures being taken more as an emotional reaction than a careful assessment of the existing reality and of citizens' real expectations.

Lack of coherence, together with an absence of coordination and planning, are the result of the inability of a political system to implement policies even in a state of emergency. Various laws and measures at the central level have been voted but seldom implemented. The inability of central government to introduce the required policies and provide the necessary resources has pushed some local governments, notably in the North, to act as substitutes but without the means or lawful authority to do so.

The degradation of traffic and environmental conditions in Italian cities urged local governments to introduce measures as early as the 1970s. The first restricted traffic zones were implemented in small and medium-sized cities, mainly to protect their historic centres. These measures (prohibiting non-residents to enter central areas, for instance) were gradually extended to larger cities in the 1980s (Milan in 1985, Bologna in 1986, Genoa in 1989, Turin in 1990), most of them being approved by referendum. Today, restricted traffic zones may be found in about 50 urban areas. They were implemented in the historic central areas of municipalities, most of them covering only a very minor part (about 2 per cent) of the municipal territory, Bologna being an exception with a limited traffic zone covering 400 ha which amounts to about 4 per cent of the municipal area. The implementation of such zones in Italian urban centres has received a great deal of generally positive publicity. However, the results are, at best, mixed. Although traffic conditions have partially improved in the zones concerned because of a dramatic decrease in traffic (between –20 per cent and –90 per cent depending on the city), they have worsened in the adjacent areas, indicating that urban congestion has not been eliminated, only displaced. Pollution does not seem to have improved either. Although no assessment of these measures has been made at national level, a study on Turin shows only a slight traffic decrease in the restricted zone (–10 per cent) and no significant change in pollution levels. This seems to be confirmed by the analysis performed by M. Tessitore (1992). Here, figures and remarks contrast sharply with the positive statements put out by the media. The mixed results are understandable for various reasons. First, the measures applied only to very limited zones in urban areas. Second, there have been many dispensations, as well as many violations due to inadequate controls. Third, the measures have not been accompanied by significant or sufficient improvements in public transport. Today, restricted traffic zones are no longer hailed as a panacea and road-pricing seems to be the new fashionable measure, although Italy is far from implementing it. At present, much is being said and nothing is being done about road-pricing.

On the planning side, urban congestion was to be dealt with by the new Urban Traffic Plans (*Piano Urbano di Traffico*) established in 1986 by the national government. These plans, which can be compared with Traffic System Management (TSM) schemes in the USA or *Plans de circulation* in France, were optional until 1992, when they were declared obligatory. Their purpose was to 'rationalize urban traffic by immediate action'. They were to be introduced in cities of over 30 000 people or where environmental problems were considered serious. No money was given to municipalities to implement such plans, however, and today they have only been approved in 30 per cent of provincial centres, the delays being due not only to the financial strain on local governments but also to resistance from many lobbies, the lack of technical expertise within the municipalities, the lack of means of control and the absence of political stability at the local level (*M & T*, 1993). Difficulties in implementing laws can be seen in other fields such as parking. Parking is a serious problem in cities mostly because no infrastructure has been built and no specific measures (such as pay-parking) have been introduced at a significant level. In 1994 in the city of Milan, for instance, only 10 000 parking spaces were paying spaces, while 300 000 on-street spaces were free. It was to remedy this problem and notably to help build parking infrastructure, that central government passed Law 122 in 1989. This law was to provide 2000 billion lire over a three year period to build car-parks featuring park-and-ride facilities in the 15 largest cities. So far, only very few have been constructed (two in Florence and two in Trieste).

With regard to pollution, specific measures to combat this were also introduced in the largest cities. These included the system of alternate registration plates by which, in periods of acute pollution, only half the vehicles may be used on any particular day. This was implemented in cities such as Naples, Bari or Palermo as a 'substitute' for restricted traffic zones and in urban areas such as Milan, Rome, Florence or Turin in specific periods only. These particular measures were insufficient to halt environmental degradation and therefore, in November 1992, a law on fighting pollution was voted for in Rome but without any means of enforcement being established. The only significant measure was the implementation of the European circular on catalytic converters in January 1993 but, to date, only 15 per cent of the urban fleet is so equipped.

The ineffectiveness of most of the measures and the government's inability to implement the various laws enacted in the last decade indicate that congestion and pollution will remain a problem for Italian cities in the near future. This seems all the more likely since public transport, which should have been developed in order to become a relevant alternative to the automobile, remains in a shambles in spite of the various changes effected since 1981.

In 1981, the situation as regards local public transport was rather confusing and delicate. Deficits had started to grow, investments were seriously needed and there was no political or legal framework within which to establish the relevant public policies to deal with these deficiencies. Law 151, enacted in the same year, was supposed to solve all these problems. Its clearly expressed objective was to rationalize local public transport operations. First, it established the region as the only authority empowered to administrate and organize public transport outside the municipalities. At local level, Law 151 thus created two authorities on separate territories, the region for the so-called extra-urban services, and the municipalities for urban services, without providing them with any instruments of cooperation and coordination. Indeed, only the largest cities had the relevant expertise and technical resources to act as 'real' authorities at urban level because they possessed large revenues, notably from the delivery of urban services such as electricity, gas and water, and they owned and controlled the large transport operators. The city of Milan, for example, controls the ATM (Azienda Trasporti Municipali) which operates three underground lines, 19 tramways, 100 bus lines and three trolley-bus lines in the city of Milan as well as in many suburban municipalities. Second, recognizing that public transport companies could not achieve a financial balance on their own, Law 151 established a subsidy programme to offset deficits as well as creating a capital programme to finance the building of infrastructure and fleet renewal. Both programmes, which formed the *Fondo Nazionale dei Trasporti* or FNT (National

Transport Fund) were for local public transport only and did not include local services run by the FS.

One of the purposes of the FNT was to help local governments and private operators eliminate their operating deficit. In 1987, it was becoming clear that this objective could not be achieved and that the FNT was insufficient to compensate for the total amount of the deficits, largely because of a miscalculation in the real operating deficits. The 1987 Finance Law then stated that the state should compensate 80 per cent of the remaining deficits accumulated in the 1982–86 period and that the remaining 20 per cent should be offset by borrowing from the banks before a new legislative framework could be implemented. In 1990, Law 403 was enacted and established that companies must achieve a financial balance before 1996. In 1992, a Presidential decree extended the 1987 Law to cover deficits accumulated between 1987 and 1991. In 1993, the FNT was abolished and integrated into the Regional Fund (a block grant), which meant that from then on transport operating deficits were to be compensated for by local governments (regions, provinces and municipalities) from their general resources. In 1994, the accumulated deficit is expected to reach 14 000 billion lire. Thus, the situation has not improved and the 1981 Law has proved ineffective.

The situation with regard to investment looks no better. Law 151 established a means of financing infrastructure and rolling stock: 75 per cent of the FNT was to be spent on fleet renewal and 25 per cent on building infrastructure. The exact sums were to be decided every year by the Finance Law. In fact, this never worked because of the government's inconsistency; it decided to fund some capital expenditure, only to withdraw the funding later, which entailed considerable delays (if not the complete postponement of infrastructure regarded as crucial to the efficient functioning of cities). Second, the amounts proposed were generally far from adequate. Finally, since 1990, previously agreed funds have been frozen because of the increasing public deficit. In 1992, however, a specific law on rapid public transport was voted for, but not implemented. Between 1981 and 1990 about 4400 billion lire (about $3.1 billion) were allocated to local public transport (Iveco, 1993), an amount which fell far short of requirements. According to a study done by Iveco (Iveco, 1990), capital needs for the urban public transport sector were about 60 000 billion lire ($40 billion), 20 000 billion lire ($13 billion) for rail nodes, 30 000 billion lire ($20 billion) for underground systems, 8000 billion lire ($5.3 billion) for fleet renewal and 1300 billion lire for park-and-ride facilities. If we exclude the 20 000 billion lire for rail nodes belonging to the urban networks of the FS (and which are therefore not included in the FNT), there is still a considerable discrepancy between the amounts injected into the FNT and the capital needs of urban areas. Government action has therefore been piecemeal at best (a specific law for Rome or for rail nodes which in all likelihood will not be fully implemented), and in some instances local governments have chosen to rely on their own resources as a substitute for governmental action and money. This was the case for the last underground line built in Italy, the M3 in Milan, which was totally funded by the city.

Local governments, central government, political parties and experts are all now convinced that Law 151 must be reformed. Two major orientations seem likely. The first would take a more comprehensive view of urban transport, while the second would take structural action to reduce operating deficits, all within a framework of decentralization with greater local autonomy and responsibility.

Considering urban transport in a more comprehensive way means parting from the modal approach which has prevailed for many decades. This modal approach, by which underground systems, trams, buses and regional railways are all dealt with separately, has led to piecemeal measures rather than an overall transport policy. Comprehensiveness would mean including the urban networks of the FS and the regional railways in urban public transport and creating policies to cover the entire sector, which has never been done before. Such a change can only be implemented if a single authority is designated to oversee each functional urban area. The double-authority system (regions and municipalities) established by Law 151 should therefore be abolished because it has produced conflicts and has impeded policy making (Lefèvre, 1991). Such a reform was introduced juridically by the vote in favour of Law 142 in June 1990 which attemps to clarify the relationships between local governments by creating a new level of government called the *città metropolitana* in the ten largest metropolitan areas. This new governmental unit (which in most cases could well be the former provinces) would control a territory representing more or less the functional territory of the metropolitan area and would be responsible for strategic planning, the environment and metropolitan networks (transport, water, and so on). It would therefore take over the planning and management of public transport from the municipalities and regions in the ten largest metropolitan areas, thus enabling better coordination in transport policies. However, as might be expected in Italy, the *città metropolitane* have not been established yet.

Structural action to reduce operating deficits is being planned. The first step would be to reduce costs, notably by cutting back on the high staffing levels in many public companies. A figure of 20 000 'redundant' employees has been estimated (Automobile Club d' Italia, 1993). The second step would be to boost revenue by increasing fares. Italian fares are considered the lowest in Europe because there is a total lack of connection between tariffs and the cost of producing services. This lack is the result of a total absence of financial responsibility on the part of public transport companies. Under the present system, subsidies come from central government and local authorities. Most urban public transport operators are publicly owned and are, in fact, part of the municipal apparatus (this is why part of the deficits can be compensated for by the profits of other municipal services). The cost of producing services is therefore unrelated to existing deficits because those who produce services (the operators) are not those who are responsible for their financial viability (local governments). Law 142 therefore proposes to establish a clear distinction between operators and municipalities by giving operators a status of autonomy (they would have to be economically viable and, if social services or social tariffs have to be established, they would be the financial responsibility of municipalities). It is believed that such a change would reduce costs, notably by forcing operators to cut duplicated services (for instance, between bus lines and tram or underground lines). Operators would have to employ a more cost-effective, private enterprise type of management.

Giving more responsibility to operators and local governments is not easy within the present Italian system because local authorities have practically no resources of their own. The bulk of local revenue comes from central government grants, most of which are allocated for specific purposes. Law 142 did nothing to change this situation because it did not address the financial aspects of the relationships between the various governmental tiers. Any reform of public transport management which does not do so is bound to be impracticable and ineffective.

CONCLUSION

Is Italy a 'blocked' society? Are Italian cities unable to produce and implement policies? The absence of policies in the sector of urban transport would seem to indicate that this is so. Indeed, the only action taken has been the introduction of piecemeal measures aimed at solving emergency situations, and most of these have proved unsuccessful. It seems that this is the only 'action' likely to be taken in Italian society today. The city is blocked both at a political and an administrative level: politically, because the necessity of forming coalition governments and the numerous parties involved in these coalitions mean that it is less risky for politicians to do nothing than to act. This lack of action is also exacerbated by confusion, and sometimes conflict, over the respective responsibilities of each institutional level (notably the municipality and the region). Nobody is responsible for area-wide public transport policy. Institutional area-wide integration, which has been achieved in most European countries, notably in Germany with the *Verkehrsverbund* model, simply does not exist in Italian cities. The situation is also blocked economically because of the state's enormous public deficit and the incapacity of the political system to establish a long-awaited fiscal reform, notably the decentralization of national taxes. Cities, provinces and regions still rely on central government money.

Of course, some cities or urban areas, such as Milan and the Lombardy region, are in a more favourable position than others. Milan was able to build its third underground line without state subsidies and the *Passante*, a long-awaited public transport interconnection infrastructure, is progressing. But so far, this remains an exception in Italy and, by European standards, Milan cannot be taken as a model for the quality of its urban transport system.

7 Great Britain: Failure of Free Market Policies

Among the countries examined in this book, Great Britain occupies a special place as the only one to have enacted a complete shift in urban public transport policy. In the 1960s and 1970s, Great Britain established a monopoly of the operation, management and planning of public transport both at the national and local level. Following the Parliamentary victory of the Conservatives in 1979, however, the Prime Minister, Margaret Thatcher, launched a new transport policy that dismantled the existing system, introducing deregulation and privatization as solutions to the worsening public transport crisis. Whereas the Reagan and Bush Administrations in the USA were unable to implement similar privatization measures because of Congressional opposition, the British Conservatives were successful in imposing their free market policies, helped by a political system which ensured them a reliable majority in Parliament and by the consistent support of the British electorate in successive national elections. Once considered the country of local government, Great Britain has become increasingly centralized, with local authorities losing most of their responsibilities and central government imposing strict control over the use of local revenues.

More than seven years have elapsed since these policies were implemented, yet they have failed to bring about most of the benefits forecast by the government when they were adopted. The decline of the public transport system, which started in the 1970s, has continued. Indeed, ridership has been falling at an even more rapid rate and this, together with worsening urban congestion, means that British cities are currently facing a dramatic transport crisis which cannot be solved whilst the government persists in maintaining existing policies and reducing public subsidies for transport.

TRENDS IN URBAN SPATIAL STRUCTURE

Great Britain was the first country in Europe to experience urbanization on a large scale. In the mid-nineteenth century, over half the population already lived in urban settlements. Today, the urbanization rate in Britain is one of the highest in the Western world, with over 90 per cent of the total population living in urban areas. More than other Europeans, the British live on the urban fringe, having deserted the traditional city centres. In this respect, British urban centres resemble their American counterparts, with urban ghettos occupying the inner cities which have been abandoned by the more affluent classes. Population decline has not been confined to city centres, however. Metropolitan areas as a whole have been experiencing a population loss, with most of them showing a decrease of over 10 per cent in the last two decades (see Table 7.1). Population growth is occurring instead in peripheral areas such as the South-east region, which has a total of 11 million people. The population here increased by 12 per cent between 1971 and 1991, and is expected to grow by another 5 per cent over the next decade (Department of Transport, or DoT, 1993a).

The decentralization of the population has been followed by the decentralization of jobs. In London, employment in the city centre has been falling for over 30 years while, in the outer metropolitan area, the number of jobs has steadily increased. The same trend can be seen in other British urban areas.

Consequently, Great Britain is now experiencing extensive suburban and ex-urban development (that is, the expansion of urbanisation to cover increasingly large areas), with a corresponding decline in traditional urban

Table 7.1 Population size (in 1000) and change (in %) in the largest urban areas, 1971–91

	1971	*1981*	*1991*	*Change 1971–81*
London (GLC area)	7 452	6 696	6 397	–14.2
West Midlands	2 793	2 646	2 511	–10.1
Greater Manchester	2 729	2 595	2 455	–10.0
Strathclyde	2 575	2 404	2 219	–13.8
West Yorkshire	2 067	2 037	1 991	–3.7
Merseyside	1 656	1 513	1 380	–16.7
South Yorkshire	1 323	1 302	1 254	–5.2
Tyne and Wear	1 212	1 143	1 090	–10.1
Bristol	427	388	372	–12.9
Cardiff	287	274	277	–3.5
Plymouth	240	244	242	+0.8

Source: Eurostat; OPCS Population Census (1991).

centres as growth poles and with most significant demographic and employment increases occurring in peripheral areas.

TRENDS IN URBAN TRANSPORT

Travel behaviour in Britain is analysed every four years or so in the National Travel Survey (NTS). However, general travel data for urban areas (over 3000 clustered inhabitants) is scarce and incomplete because it does not include journeys of under one mile, and these are especially important in urban areas. The 1985/86 NTS (DoT, 1988), for instance, calculates an average of 1.9 journeys per person per day. Separate data for urban areas has not yet been published from the 1989/91 NTS (DoT, 1993c). It is possible, however, to make a comparison by using total population figures since only 11 per cent of the population is defined as 'rural'. The average British citizen made 2.6 trips a day in 1975/76, about 2.8 trips in 1985/86 and 3.0 trips in 1989/91, indicating a slight increase in mobility. This increase does not, however, apply equally to all types of journey. As in other countries, the journey to work has consistently declined as a proportion of total travel, from 46 per cent of all trips in 1965 to 33 per cent today, whilst leisure and personal journeys have grown in relative importance (38 per cent and 29 per cent respectively in the 1989/91 NTS). Not only are the British making more trips; they are also making longer trips. In urban areas, the average distance for a trip increased by 9 per cent, from 11.4 to 12.5 km between the last two NTSs, with London showing the greatest growth. With an average distance for commuting trips in central London of 24 km, the Londoner travels much more than other urban residents who average 11.2 km in metropolitan areas other than London, and 9.6 km in smaller urban areas. Journey patterns in urban areas nevertheless remain unclear. Since there is no information on travel destinations in Britain currently available, it is difficult to know whether mobility is following trends similar to those in France and the USA. However, considering the similarities in urbanisation patterns, it is likely that it is following the same general trend of increased suburb-to-suburb travel.

The car has become an increasingly dominant form of transport and its use has increased dramatically over the last decades, as shown in Tables 7.2 and 7.3. In all urban areas, but particularly in London, the car is the only mode that shows an increase, largely to the detriment of all other modes. A decline in cycling is particularly noticeable, but the importance of public transport and walking has also declined. Table 7.4 for London illustrates this development.

Table 7.2 Change in modal split, London, 1975–91 (% journeys and % mileage)

	1975–76	*1985–86*	*1989–91*	*1975–91 % change*
% mileage				
Car	63.2	68.2	68.9	+9
Public transport	27.8	24.2	24.7	–11
Two wheelers	2.5	1.8	1.2	–52
Walk	6.5	5.7	5.2	–20
% journeys				
Car	41	44.3	47.8	+16
Public transport	20	17.3	17.0	–15
Two wheelers	3	2.8	1.7	–43
Walk	35	35.0	32.7	–6

Source: DoT (1993a).

Table 7.3 Change in modal split, urban areas, 1985–91 (% of mileage)

	1985–86	*1989–91*	*1985–91 % change*
Car	74	77.5	+5
Public transport	16	15.0	–6
Two wheelers	5	3.5	–30
Walk	5	4.0	–20

Note: journeys under one mile are included but, since most journeys walked are short, the importance of walking is less significant.
Source: National Travel Surveys, corresponding years.

A decline in public transport ridership has been one of the major features of urban transport in Great Britain over the last two decades and is all the more striking when compared to ridership increases in most European countries during the same period. It should, however, be borne in mind that unlike countries such as France, Great Britain already had a high level of public transport use in terms of trips per inhabitant. Table 7.5 shows details of ridership trends for each type of public transport from 1975 to 1992. Buses were the most severely affected mode, with a 20 per cent drop in use both in London and other metropolitan areas during the decade prior to deregulation (1975–1985). Since 1986, ridership decline has been much more dramatic in the deregulated

Table 7.4 Main mode of travel to work by area of residence in London, 1971–91 (%)

Residents	*Mode*	*1971*	*1981*	*1991*
Inner London	Public transport	54	52	51
	Car	18	26	30
	Walk	19	17	14
Outer London	Public transport	43	37	35
	Car	34	46	52
	Walk	15	12	9
All London	Public transport	47	42	40
	Car	27	39	45
	Walk	17	14	10

Source: DoT (1993a).

Table 7.5 Changes in local public transport usage in metropolitan areas, 1975–92 (in millions of passenger journeys)

	1975	*1980*	*1985/86*	*1989/90*	*1991/92*	*1992/93*
LBL	1 455.0	1 183.0	1 152.0	1 188.0	1 149.0	1 127.0
PTE	2 313.0	1 973.0	2 068.0	1 648.0	1 481.0	
BR*	13.4	13.0	13.0	15.2	14.3	13.6
LUL	601.0	559.0	732.0	765.0	751.0	728.0
Strathclyde	12.1	6.8	13.7	13.5	13.6	13.5
Tyne and Wear		14.1**	59.1	45.5	40.6	38.9
DLR				8.5	7.9	6.9

*British Rail Network South East in billions passenger-km.
**1981 data.
LBL: London Bus Ltd.
PTE: Passenger Transport Executive.
LUL: London Underground Ltd.
DLR: Dockland Light Railway.
Source: DoT (1993b).

metropolitan counties (–28 per cent) than in London (–2 per cent: see Table 7.8).

Trends in rail travel have not followed the same pattern, however. In London, underground traffic, in terms of the number of journeys made, grew steadily between 1982 and 1989 (the number of journeys rising from 498 to 815 million), but started to fall in 1989 (from 815 to 728 millions in 1992/93). British Rail's Network South East, which serves the Greater London area, followed the same trend as the underground, with an increase in passenger kilometres between 1983 and 1989 and a decline from then on. On average, however, passenger kilometres increased by 9 per cent between 1983 and 1993. Other metropolitan areas show mixed trends. In Strathclyde (the Glasgow region), underground traffic grew in terms of passenger journeys from 1982 to 1986 and in passenger kilometres from 1982 to 1989, and now seems to have stabilized, whereas in Tyne and Wear (the Newcastle area), the number of journeys has fallen steadily since deregulation whilst passenger kilometres have fluctuated (310 million in 1984, 319 in 1989/90 but 271 in 1992/93).

In common with other European countries, Great Britain has experienced skyrocketing car ownership (see Table 7.6.) and use, although to a lesser extent than France or Germany.

In the last two decades, the number of cars has risen by 60 per cent, from 12.5 million in 1975 to 20.1 million in 1992, whilst the number of private cars per 1000 people rose from 251 in 1975 to 386 in 1991. Not only have more and more British families become car owners, but more cars are being bought by each household. Multi-car ownership has increased steadily and its level doubled between 1975 and 1991, from 10 per cent to 21 per cent. However, the percentage of households without access to a car is still significant (35 per cent), and larger than in most European countries. Figures in London show the same trend, but slightly below the national urban average.

Table 7.6 Household car ownership in urban areas, 1976–91, and in London (1971–91) (%)

	Urban areas			*All areas*			*London (GLC)*		
	1976	*1986*	*1991*	*1976*	*1986*	*1991*	*1971*	*1981*	*1991*
0 car	48	40	35	45	38	33	54	45	40
1 car	43	44	44	45	44	44	38	42	41
2 or +	10	16	21	11	17	23	8	14	19
per 1000	237	314	370	251	329	386			

Sources: DoT (1988, 1993c).

Largely as a result of the increase in car ownership, automobile traffic has grown significantly in the last decade. In urban areas, car-driver mileage increased by 28 per cent and car-passenger mileage by 23 per cent between 1985 and 1991. Figures for London show similar trends since traffic has increased by an average 2 per cent a year since 1971 (DoT, 1993a).

As elsewhere, traffic growth has necessitated road construction. Although this reached significant proportions in the 1960s and 1970s when the bulk of construction occurred, road building has been less extensive than in the USA, France or Germany. In 1960, there were only 153 km of motorway in the UK, but by 1975, this had increased to 1900 km, and by 1992 to 3076 km. Motorway construction was the fastest growing of all road building activities between 1982 and 1992. In this decade, the motorway network increased by 17 per cent, while the total road network grew by only 5 per cent. However, most of the construction did not take place in urban areas (there were 13 km of urban motorway by 1960, 90 km by 1975 and only 132 km by 1985: Hass-Klau, 1990). Since then, new construction has been minimal.

Generally speaking, urban road network conditions in the largest areas are considered poor (pot holes, poor maintenance, and so on). In the Greater London area there are fewer than 250 km of motorway. Only one major road has been completed in recent years (the M25 in 1986) and the present network is regarded as encompassing narrow multi-purpose roads, incomplete radials and only one partial orbital road (Bayliss, 1991).

Public transport conditions are also very bad. Virtually no significant infrastucture has been built over the last 10 years. London excepted, most of the ridership is still carried by bus. In metropolitan areas (London and the six metropolitan counties), bus mileage has increased significantly in the last two decades (+17 per cent between 1975 and 1993) but, London apart, growth mainly occurred since 1986 (deregulation year), and the current bus network cannot easily be compared with that which preceded deregulation. No significant rail construction has been undertaken, although many projects have been drawn up. Until the early 1980s, there were only two underground rail systems in Great Britain (one in London, the other in Glasgow). The Glasgow underground line is just 11 km long and, although extensively modernized at the end of the 1970s, it has never been extended. In London, only one new line, the Jubilee, was built in the early 1980s and recent extensions to the whole network amount to a mere 10 km. In 1975, the London underground was 381 km long, yet by 1993 only 13 km had been added to form a 394 km system.

In the 1980s, two LRT systems were built. The first opened in Tyne and Wear in 1981 and is now 59 km long. The second, the London Docklands (DLR) LRT, opened in 1988 and is now about 20 km long. Since then, many projects have been drawn up in more than 20 urban areas, but so far only two have been completed: the Greater Manchester link, which opened in 1992 on 31 km of existing British Rail (BR) track, and the 8 km Sheffield Supertram line, which went into operation in March 1994.

URBAN TRANSPORT PROBLEMS

Transport problems in British cities are not so very different from those besetting other European or American towns. Road congestion and air pollution have started to plague urban areas. Lack of investment in public transport has curtailed service supply, leading to congestion in this sector, too. Before deregulation, skyrocketing subsidies proved insufficient to offset the decline in passengers. By international standards, road safety seems to be British urban transport's sole success story, and British roads are among the safest in Europe.

We will now deal with the various sets of transport problems in turn, taking first the social and environmental effects of car use (that is to say, pollution, congestion and safety), and then the crisis in public transport.

The social and environmental problems created by car use

With the exception of road safety, nation-wide data on the major problems created by car use (that is, congestion and air pollution in urban areas) is fairly difficult to obtain. As in other countries, data is scarce, inconsistent, fragmented, and mostly specific to only a few urban areas, making diagnosis a rather uneasy task.

Air pollution

Motor vehicles have become increasingly responsible for air pollution over the last ten years. This becomes clear if we look at the part they play in the creation of two major air pollutants, NOx and CO. In 1982, motor vehicles produced 38 per cent of the NOx and 83 per cent of the CO found in the air, whilst by 1991 these figures had risen to 52 per cent and 89 per cent respectively. In London, where air pollution data is available from 1981 to 1992, nitrogen dioxide was relatively stable whereas lead pollution declined sharply by 75 per cent. Ozone concentration, on the other hand, increased in central London and exceeded the Department of the Environment (DoE) threshold for poor air quality for four years during the same period. Air pollution is becoming an increasingly important social issue, as has been confirmed by various polls (Jones, 1991).

Road congestion

Traffic growth has remained constant over the last two decades. In 1975, about 165 billion vehicle-km of car travel were undertaken. By 1992, this had doubled to 335 billion and the DoT forecasts an increase of between 83 per cent and 142 per cent by the year 2025, as a result of increasing car ownership and use. Although the average speed of car travel increased by a steady 13 per cent between 1975 and 1991, growth was less for shorter (under 16 km) journeys (+7 per cent only). However, since shorter journeys are made in both rural and urban areas, this slower increase in speed may be due to increasing congestion in towns. The hypothesis of increasing traffic congestion in urban areas is confirmed by the situation in London. In the Greater London area, the average traffic speed has been steadily declining since 1968. The relevant data for the whole area (excluding the newly built M25 which is already badly congested) indicates a general decline in speed at all times of day and for all parts of the region.

Road safety

The UK, Norway and Sweden have the best traffic safety records in Europe. This does not mean that safety on the roads is not a problem but, in comparative terms, the British situation is exemplary. In 1975, there were 6366 fatalities and 83 488 injuries on British roads. Road accident casualties then declined steadily to 4229 deaths and 53 474 injuries in 1992, representing decreases of 34 per cent and 36 per cent. Data specific to urban areas is not available for the corresponding periods except for London and the West Midlands (Birmingham area). In London, there was a 41 per cent decline in fatalities and a 10 per cent decline in serious injuries between 1981 and 1992. In the West Midlands there was a decline of about 30 per cent in both fatalities and serious injuries between 1977 and 1992. There is obviously still much room for improvement, since about 70 per cent of those interviewed consider road safety a major issue (Jones, 1991), and the government has recently launched initiatives to reduce road traffic injuries by one-third by the year 2000.

Public transport decay

The absence of investment in public transport, especially in rail infrastructure, has largely contributed to the increasing congestion of public transport networks. The situation is most serious in London where the underground rail system and the BR regional network run services over more than 3000 km of tracks and serve over 1000 stations. Together they cover over 20 billion passenger-km a year. However, much of the system is old; modernization and renewal have, for many years, been inadequate. The recent upsurge in traffic demand has put additional strains on the networks and has resulted in overcrowding at peak hours and deterioration in the quality of the service (Bayliss, 1991). These shortcomings are particularly acute in and around the centre, and in the rapidly growing Docklands area. Figure 7.1 illustrates this situation.

In the ten years leading up to 1992, the number of people entering Central London by BR during the morning peak period increased by 3 per cent (compared to 24 per cent between 1983 and 1988) and on London Underground by 19 per cent (a slight decline since 1988). At the same time, the number of people travelling by bus declined by 38 per cent, and by car by 24 per cent, thus putting more strain on rail services.

Before 1986, central and local governments injected tremendous amounts of money into the system to cover increasing deficits, but with no substantial results. According to the DoT, operating deficits were also largely due to a drastic growth in operating costs (+70 per cent per vehicle kilometre between 1970 and 1983 for London Transport, +30 per cent for metropolitan counties). This heavy increase in revenue support (from 0.4 pence per stage (local or less than 30 miles) passenger journey at 1980 prices in 1972 to 7.9 pence in 1982) meant drastic reforms were called for. Deregulation and privatization were thus introduced in an attempt to solve this problem.

TRANSPORT POLICIES

In 1979, a Conservative government led by Margaret Thatcher came to power. Its policy was to reduce public expenditure and to favour the involvement of the private sector in the economy. These objectives implied a complete redefinition of the role of the public sector and notably of local governments, since they were considered to be among the biggest spenders of public money. The fact that most large urban areas were controlled by the Labour Party only exacerbated matters. From an economic

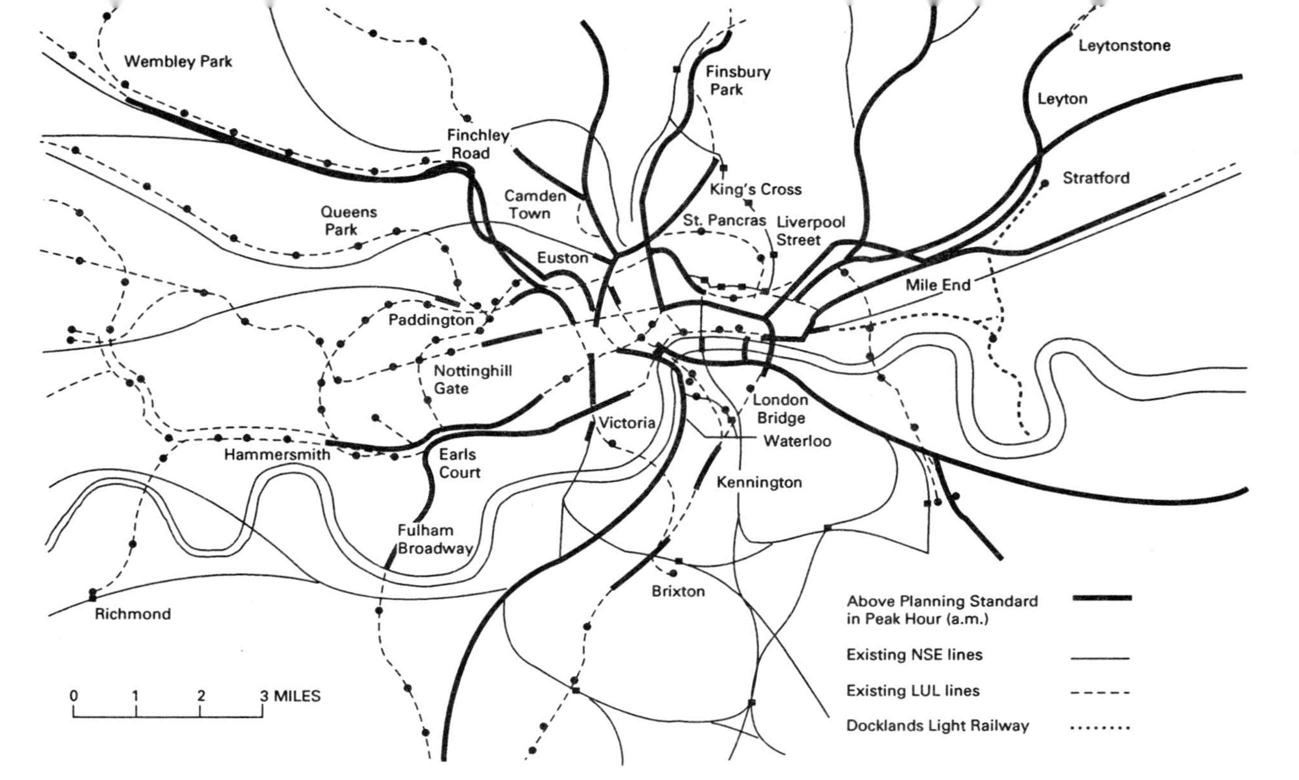

Figure 7.1 Saturated railway lines in the London area

Source: Bayliss (1991).

point of view, deregulation and privatization were considered the most appropriate ways to increase efficiency, whilst from a political point of view, the reduction of local autonomy and local government powers (that is, centralization) were the major policies. An example of such policy was rate-capping, the central limitation of local government fiscal powers, which was introduced in 1984 in order to prevent some local authorities using their fiscal revenue to compensate for the diminution of central government grants.

Energy (gas, water and electricity), telecommunications and air transport were sectors where it was decided to introduce deregulation and privatization. Transport, and specifically public transport, were targets for similar juridical and economic changes. Long-distance bus services (coaches) were deregulated in 1980, followed in 1986 by local bus services (stage). We will focus first on public transport policies in the last two decades, then analyse the situation regarding private transport and conclude by discussing alternatives for achieving a more balanced transport system in British cities in the future.

Public transport in the deregulation years

Before detailing current policies and results in the field of public transport, we will describe the organization of the system prior to deregulation and the major deficiencies criticized by Mrs Thatcher's government.

The 1968 Transport Act established a framework for public transport activities in British urban areas outside London (White, 1987). This Act created the National Bus Company (NBC) in England and Wales and the Scottish Bus Group (SBG) in Scotland. Both companies were publicly owned by central government and were holdings with several regional subsidiaries. Although essentially operating national or regional services, most small and medium-sized cities were served by these companies prior to deregulation. In about 50 cities, public transport was operated by municipal undertakings.

In four of the largest urban areas (the West Midlands, Merseyside, Greater Manchester, and Tyne and Wear), the 1968 Act established PTAs, joint-boards grouping the various districts in the area. Public transport operation and planning in these areas was carried out by Passenger Transport Executives (PTEs), playing a double role of transport operator and technical consultant (planning, transport studies, and so on) and controlled by the PTA. The PTEs had the monopoly of operations in these areas which included not only buses, but also the local BR services. Fares were determined by the PTA which also distributed operating subsidies and capital grants.

In 1972, following the Local Government Act, the PTA and PTE functions were transferred to the newly created 'metropolitan counties' in the six largest areas (the four previously mentioned, plus South Yorkshire and West Yorkshire). Gradually, these metropolitan county councils established completely unified public transport networks. Fares were coordinated and integrated in most urban areas during the 1970s, as well as bus and rail services. When the Conservative government took over in 1979, public transport in urban areas was under a complete public monopoly. A policy of low fares, mostly in metropolitan counties, had led to increasingly large deficits, amply compensated for by local governments from their own revenues, and from subsidies automatically given by the state.

The situation in London was rather similar. The Greater London Council (GLC) had been established in 1963 but was only given responsibility for public transport in 1970. Unlike the other metropolitan areas, it did not, however, control the BR network serving its region. Fare integration in London was introduced in 1983, but only concerned London Transport services (bus and underground).

The Thatcher government declared that public subsidies for public transport operation were incompatible with its objective of reducing public spending. In the White Paper, *Buses* (DoT, 1984), it clearly stated that 'the cost of subsidizing public transport is now unacceptable', referring to the fact that over the country as a whole, revenue support had gone up from £10 million in 1972 to £520 million in 1982, a 13-fold increase in real terms. For the Conservatives, the organizational and financial policies established over the years in the field of public transport were fully responsible for the poor situation in which it found itself. They pointed in particular to the 'very highly restrictive licensing system', the overprotection of public operators through restrictive regulation and the cross-subsidization of services, adding that the best parts of the market had been neglected. Urban areas were clearly the Government's main targets because the 'big subsidies' were being paid in London and in the metropolitan counties, accounting for 80 per cent of total revenue support, while comprising only 40 per cent of the population of England. The policies of the larger urban areas were also severely criticized for holding down fares to artificially low levels and for maintaining services which were unjustified by the low level of demand. Competition and deregulation were judged to be the most appropriate solutions since it was thought they would enhance initiatives and efficiency.

Drastic measures were soon to be enacted. Even before the publication of *Buses*, the government had introduced deregulation for express bus services (over 30 miles) in the 1980 Transport Act. In addition, in three 'trial areas', local bus services (under 30 miles) were also deregulated. Since it considered that the effects of the 1980 Act had been positive, the Government introduced a general deregulation policy for bus transport.

The deregulation of bus transport was viewed as only the first step towards the total deregulation of public transport in Great Britain, and the government has always made clear its intention to deregulate and privatize the railway and London's public transport in the near future.

In *Buses*, the government was very explicit regarding its expectations of the reform: better services, more innovations, lower fares, ridership increase and a dramatic fall in public subsidies for public transport. At the same time, the Conservatives seemed ready to curtail the largest public money spenders, the metropolitan county councils and the GLC. In 1984, the government had already relieved the GLC of responsibility for public transport and in 1985, by approving the Local Government Act, the British Parliament eliminated the only directly elected metropolitan authorities in the Western world. This had a profound effect on the organization of public transport in the largest urban areas. PTAs were transformed into joint authorities consisting of members nominated by each constituent district council. Public transport was no longer under the control of directly elected local goverments.

The 1985 Transport Act introduced major changes in three areas: market entry, transport operations and the role of local governments. The traffic licensing system established decades previously was abolished. Anyone able to satisfy certain quality and safety standards may now enter the market. Registration of services is still required and must be carried out at least 42 days before operating begins, as must deregistration.

To ensure fair and open competition, several changes were introduced: (1) in the bus industry, public undertakings (NBC, SBG, PTEs, Munis) were to be dismantled and privatized; (2) in the PTAs and PTEs, bus operations were to be transferred to independent companies known as Passenger Transport Companies (PTCs); (3) cross-subsidization between profitable and unprofitable routes was forbidden; (4) cooperation between operators was no longer obligatory; and (5) access to public transport facilities (for example, bus stops) was to be given to all operators on an equal basis.

The role of local government was transformed. Before deregulation, local authorities were in complete charge of public transport operations, management and planning. After deregulation, their role was much less important. By opening the market and introducing competition, the 1985 Transport Act established a system based on two types of service: commercial services (that is, those which would be considered profitable and could be run without being subsidized), and social services, which would be non-profitmaking and subsidized on the grounds that they were to be established or maintained for social reasons. About 85 per cent of services were registered as commercial, with local goverments taking responsibility for social services only (that is, a small part of the network). They were to subsidize these after a tendering process. Local authorities were also responsible for concessionary travel, which represented about 25 per cent of all trips in some areas. In metropolitan counties, local governments (PTAs) lost control of commercial routes but remained in charge of rail services (planning, fares, level of services) since rail was not deregulated.

Not surprisingly, the role of local governments has become less significant as a result of deregulation and privatization, since their control over public transport is now largely limited to social services and rail routes in metropolitan areas.

Deregulation began on 26 October 1986 with a three-month 'freeze' period in which services could not be altered once registered. Eight years have elapsed since 1986, long enough to allow an analysis of the impact of the reforms introduced by the 1985 Act.

In the bus industry, all NBC subsidiaries were sold between 1986 and 1988 and the same process was completed in 1991 for the SBG. Operators owned by the PTAs or by other local authorities were redefined as 'arm's length' PTCs. All of them were dismantled and privatized in the metropolitan counties. In other urban areas, almost all of them have now been privatized, although about 20 operators remain under the control of district councils (in Reading and Cardiff, for instance).

Although there have been quite a few newcomers, competition has not, on the whole, been intense, except in specific areas such as Strathclyde, the Southern area of Greater Manchester, Tyne and Wear and Oxford (Perett *et al.*, 1989). In general the large, previously public, undertakings have continued to control the bulk of operations. Bus supply (in vehicle km) has increased substantially, as shown in Table 7.7. As already mentioned, the commercially registered proportion of service kilometres has been both high (about 85 per cent) and stable (Table 7.9). However, fares have increased drastically since 1985–86 (Table 7.10), mainly in metropolitan areas, albeit from a much lower level than in other areas. As previously stated, ridership fell significantly due to fare increases and to the considerable disruption of services which occurred in the first two years of deregulation (Table 7.8).

Deregulation brought substantial benefits as regards levels of public subsidy. Total public transport subsidy outside London declined by a substantial 56 per cent between 1986 and 1992. In view of the high level of

Table 7.7 Percentage change in local bus-kilometres run, 1985–86 to 1991–92

Area	%
Great Britain	+20
Metropolitan counties	+15
Shires (England)	+22
Scotland	+25
Wales	+26
London	+15

Source: White (1993).

Table 7.8 Percentage change in local bus passenger journeys by area, 1985–86 to 1991–92 (%)

Area	%
Great Britain (outside London)	–22
Metropolitan counties	–28
Shires (England)	–16
Scotland	–17
Wales	–18
London	–0.3

Source: White (1993).

Table 7.9 Types of services registered, 1991–92 (in millions vehicle kilometres and % of commercial services)

	Commercial	*Subsidized*	*Total*	%
English metro areas	568	93	661	86
English shire counties	850	187	1 037	82
Scotland	304	51	355	86
Wales	94	26	120	78
All outside London	1 816	357	2 173	84

Source: White (1993).

Table 7.10 Local bus services, fare indices by area, 1984 to 1991–92 (1985 = 100)

Area	*1984*	*1985–86*	*1988–89*	*1991–92*
London	91.6	101.7	125.3	167.0
English metro areas	98.6	100.4	146.6	197.0
English shires counties	94.8	101.2	117.6	151.4
England	95.3	101.1	127.0	165.8
Scotland	98.3	100.2	112.2	136.6
Wales	96.3	101.1	n.a	n.a
All Great Britain	95.8	100.9	124.3	160.8
All outside London	96.5	100.8	124.2	159.8
All outside London and metro areas	95.7	101.0	115.9	147.6
Retail Price Index	94.3	101.2	115.1	142.5

n.a. = not available.
Source: DoT (1993b).

commercially registered services (that is, services operated without subsidies), a sharper reduction might have been expected, but this was partly offset by the significant growth in concessionary travel (special groups such as senior citizens paying reduced fares: White, 1993). Operating costs were also reduced. Unit operating costs per bus km have declined sharply, by about one-third since 1986 for Britain as a whole and by 40 per cent for the six metropolitan areas. This reduction is due to many factors: a 50 per cent fall in real fuel costs per litre, a diminution in real weekly earnings per employee, an increase in staff productivity resulting from the introduction of more flexible operations, a reduction in maintenance and engineering staff and the rapid growth of the minibus (the total cost of minibuses being about 65 per cent less than for full sized vehicles, largely because of lower driver wage rate: see White, 1993). Finally, some innovations, principally the development of mini- and midibuses, have occurred since 1986.

Nonetheless, deregulation did not bring about most of the positive effects forecast by the government. Indeed, some adverse effects can now be seen. Competition did not yield fare reductions nor an increase in ridership. Public spending has been significantly reduced but to the detriment of overall mobility (equality of access is likely to have been affected) and, as some experts have suggested, also to the detriment of the long-term profitability of the industry and fleet renewal. Profitability remains poor, at about 2 per cent in 1990–91 (White, 1993). Fleet renewal has been low, which signifies an ageing fleet, probably reducing service quality and rising maintenance costs in the near future.

These rather negative results should have led to prudence in extending deregulation. Nevertheless, the government has confirmed its intention to continue with it in London, and the national rail services are now being deregulated.

For the Conservatives, London was clearly the major area where the previous public transport organisation had proved inefficient. The GLC's policies of the early 1980s (notably the famous 'Fares Fair Policy': in 1981, the GLC had launched a policy of low fares on its network, which resulted not only in an increasing deficit but also in a dramatic ridership increase and was abandoned in 1984, due to the opposition of some London boroughs) were definitely not in compliance with the government's objective of public expenditure reduction. In 1984, the state took over the control of public transport from the GLC (which was abolished one year later). It separated London Transport into three distinct companies: London Underground Ltd, London Bus Ltd (LBL) and the Docklands Light Railway (which was to be in charge of operating the new rail system to be developed in the Docklands area). Although the London bus network was to be deregulated and privatized at a later date than networks elsewhere in the country, competitive tendering was introduced in 1986 in anticipation of such a situation, with a view to improving LBL's productivity. LBL was also divided into 12 autonomous subsidiaries. By late 1993, about 50 per cent of the old bus network had been tendered out, with both LBL and independent companies successful in gaining contracts. The other half of the network is operated under a direct net subsidy contract with LBL companies.

Such measures have had some positive effects in London. Costs have decreased dramatically, the average cost per bus kilometre having dropped 33 per cent between 1985 and 1993. Operating subsidies have also declined sharply from 42 per cent of the total operating cost in 1985 to 22 per cent in 1993. Fare increases (+25 per cent in real fare level) have also played a role in these favourable results. However, bus usage has fallen with a 2 per cent decrease in passenger trips and a 6 per cent decrease in passenger-kilometres between 1985 and 1993.

As far as the balance between usage and cost is concerned, London's results are much better than those of deregulated areas. Nevertheless, in 1994, the government confirmed its intention to extend deregulation to London in the coming years.

No solution to the dominance of the car

Traffic problems have increased significantly in urban areas as a result both of the considerable rise in car ownership and use, and the lack of a major road building programme, the latter partly due to community opposition ('Homes before Roads') in the last decades. Traffic management has long been considered the only policy capable of dealing with automobile traffic congestion, except in the 1960s, when the bulk of road building was carried out. One-way street schemes and on-street parking restrictions were implemented to extend road capacity to its maximum. Capacity can now no longer be increased and policies are being oriented towards the reduction of automobile use. The recent proposal of a network of 'Red Routes' in the London area, a 480 km road network as of the year 2000, with intensive parking controls and delivery and stopping restrictions, coupled with further traffic management measures and tougher enforcement of traffic, is illustrative of this change. Projects involving high technology, such as electronic vehicle navigation systems, have also been suggested to reduce automobile congestion.

However, traffic problems have not been, and will not be, solved by traffic management and high technology. Opposition to road construction is as strong today as it was decades ago. Most road building schemes proposed for the London area in recent years have been withdrawn because of public pressure. Not only political and social obstacles but financial constraints are considerable, too. The government has always been strongly in favour of involving the private sector in the financing of infrastucture. So far, experiences of private financing have been rare in spite of the Dartford–Thurrock bridge scheme and the optimism of the DoT in 'bringing in private finance' (DoT, 1989). It is now widely accepted that new infrastructure is not the solution to automobile congestion (Younes, 1993). Road-pricing, the new leitmotiv of various interest groups (Association of London Authorities, Chartered Institute of Transport, the Institute of Civil Engineers, and so on), is often debated, but no experiments have so far been introduced. There is strong resistance to road-pricing in Britain (Jones, 1991) which is generally considered a 'policy of last resort'.

For all these reasons, urban congestion will remain a problem in British cities for the next decade. Most experts now believe that a shift towards a more balanced transport system (that is, measures more favourable to public transport and non-motorized means of transport, such as cycling or walking) will need to be seriously considered in order to reduce environmental pollution and traffic congestion. But can this goal be achieved in Britain today?

Policies which favour alternative means of transport

Separating people from their cars is a well-nigh impossible task. It would require a considerable increase in public transport investment and the implementation of policies which favour public transport and alternative modes, neither of which seems very likely in the near future.

The saturation of public transport networks and the degree of deterioration of many facilities indicate a severe lack of investment in public transport. While public transport accounted for about 32 per cent of total local public transport capital expenditure in 1981/82, it only represented 13 per cent of such expenditure in 1992/93. This sharp decline has not been steady over the years and is obviously the result of the 1986 deregulation and privatization measures. However, since 1986 no significant sums of private money have been invested in public transport infrastructure, and in London this is in urgent need of improvement. Building new underground lines and extending others, and upgrading London Transport and BR networks is of crucial importance for the capital. This has been acknowledged by the proposed extension of the Jubilee Line to serve the Docklands area, the construction of an express connection between Heathrow Airport and Central London (the Heathrow Express), the BR programme for upgrading the former Network South East and the planned construction of an LRT from London to Croydon. Some of these projects, namely the Heathrow Express and the Croydon line, are being financed in part by the private sector. However, in other urban areas, progress is slow since many projects, including the Bristol LRT, have been delayed or postponed because of a lack of available funds. Generally speaking, private sector involvement in the financing of public transport infrastructure has been very limited (3 per cent, that is £7 million out of £240 for the Sheffield LRT, and £5 million (also about 3 per cent) for the Greater Manchester Metrolink).

Policies in favour of public transport and alternative modes, and those which aim at restraining automobile use, are not seriously contemplated. British people still consider walking and cycling as dangerous and the relevant facilities are either absent or neglected (Tolley, 1990). In the early 1980s, the GLC launched a construction programme for 1600 km of cycle ways and in 1986 the DoT proposed urban cycle route network projects in a number of cities (Bedford, Cambridge, Canterbury, Nottingham and Stockton). Spending 1 per cent of total capital expenditure on alternative modes was considered desirable for local governments, but because of budget cuts this figure was never reached and most of the projects were never fully pursued (McClintock, 1990).

However, the government recently recognized the necessity for a more comprehensive assessment of transport needs in order to allocate funds to local highway and passenger transport authorities. To do so, in 1993 the DoT adopted the so-called 'package approach'. A 1993 government circular encouraged local authorities to bid jointly for subsidies for both public transport and roads. It gave local governments greater flexibility in switching resources between the various modes of transport in urban areas. While it is too soon to know the results of such an initiative, and although funding mechanisms still remain largely geared to road schemes, the 'package approach' may well mark a significant change in the way central government and other public authorities view the question of urban transport in the near future.

CONCLUSION

Compared to recent events, the history of urban transport in Great Britain is one of inconsistent policies allied to a firm belief on the part of central government that the private sector would compensate for the withdrawal of public money from that domain. This did not happen, and British cities are now suffering from a lack of investment in public transport as well as in road networks. In view of the length of time required to build such infrastructure, this situation will undoubtedly continue into the near future. The only domain where a consistent public policy (deregulation and privatization) has been exercised is in the public transport sector, but it did not bring the success expected of it. In the light of current results in the London area, it could be argued that London is not a bad example of a compromise between the necessity to reduce costs and to increase productivity on the one hand, and the need to prevent ridership decline on the other. However, following such an example would very likely require a political change within central government, which may very well occur in the near future.

8 Eastern Europe: Transport Impacts of Political Revolution

The countries of Central and Eastern Europe have been undergoing profound political, social, and economic changes over the past five years. The overthrow of Communist dictatorships throughout the region from 1989 to 1991 unleashed a series of revolutions, affecting virtually every aspect of life. Among former Socialist countries, East Germany (the German Democratic Republic) experienced the most sudden transformation, thanks to its reunification with West Germany in 1990. The developments in East Germany will be discussed in this chapter because East Germany was an integral part of the political and economic system in Eastern Europe until 1989. Moreover, the situation there is useful for predicting future developments in other former Eastern Bloc countries. The rest of the chapter focuses on Poland, Hungary and Czechoslovakia (since 1993, this has split to become the Czech Republic and Slovakia). Of course, political and economic changes in Bulgaria, Romania, the former Yugoslavia and the former Soviet Union have been at least as important, but impacts on urban transport there have not yet been as large, and data for those countries are less available and less reliable than for the countries included in this chapter.

The most significant change in urban transport in Eastern Europe has been the shift from primary reliance on public transport towards ever greater ownership and use of automobiles. That modal shift has caused enormous social and environmental problems in Eastern European cities: increased traffic congestion, pollution, noise, accidents and mobility problems for the poor. Those problems have been particularly serious because of the suddenness of the increase in car use and the lack of appropriate transport policies and financing in Eastern Europe. Since the overthrow of Communism, public policies in most countries have permitted and even encouraged increased automobile use. Unless more is done to control car use and improve public transport, irreparable damage will be done to Eastern European cities and their transport systems. Indeed, the transport crisis in Eastern Europe is more extreme than in Western Europe or North America, where the problems of automobile use have arisen more gradually and where far more resources are available to deal with transport problems.

URBAN TRANSPORT UNDER SOCIALISM

Even during the last two decades of Socialism, car ownership and use had begun to grow considerably in Central and Eastern Europe. From 1970 to 1988, car ownership per 1000 inhabitants increased from 15 to 119 in Poland, from 22 to 163 in Hungary, from 46 to 196 in Czechoslovakia, and from 65 to 225 in East Germany (see Table 8.1). On the eve of Socialism's demise, the German Democratic Republic had achieved the highest level of car ownership of any of the world's socialist countries, with five times the car ownership rate in the Soviet Union (225 as against 42).

One might expect that such increases in car ownership would have led to dramatic reductions in public transport use and to corresponding declines in public transport's share of modal split. On the contrary, public transport use actually increased. As shown in Table 8.2, ridership grew in

Table 8.1 Trends in car ownership in Eastern and Central Europe, 1970–2000 (cars per 1000 population)

	Poland	*Hungary*	*Czechoslovakia*	*Czech Republic*	*Slovakia*	*East Germany*
1970	15	22	46			65
1980	67	95	152			150
1985	98	135	180			199
1988	119	163	196	216	158	225
1990	138	189	211			296
1992	169	217	219	241	176	415
2000	210	260		306	200	460

Note: Even before the division of Czechoslovakia on 1 January 1993, the Czech and Slovak Republics existed as administrative districts and had exactly the same geographic borders as currently.
Sources: Ministries of Transport of Poland, Hungary, Czechoslovakia, Slovakia, the Czech Republic and Germany; Klofac (1989); Lijewski (1987); Mitric (1994); Polish Central Statistical Office (1982 to 1994); Pasquay and Monigl (1992); Prague Institute of Transportation Engineering (1994); Pucher (1993); Rataj (1988), and Suchorzewski (1993).

Table 8.2 Trends in public transport ridership in Eastern and Central Europe, 1980–92 (millions of trips each year with percentage of 1980 level in parentheses)

	1980	*1985*	*1988*	*1990*	*1992*
Poland	7 370	8 964	8 965	7 264	6 030
	(100)	(122)	(122)	(99)	(82)
Hungary	3 105	3 524	3 581	3 134	2 687
	(100)	(113)	(115)	(101)	(86)
Czech Republic	2 712	2 983	3 119	3 122	3 309
	(100)	(110)	(115)	(115)	(122)
East Germany	3 435	3 524	3 531	2 802	1 567
	(100)	(102)	(103)	(82)	(46)

Sources: Ministries of Transport of Poland, Hungary, Czechoslovakia, Slovakia, the Czech Republic and Germany; Klofac (1989); Lijewski (1987); Mitric (1994); Polish Association of Urban Public Transport (1994); Polish Central Statistical Office (1982 to 1994); Pasquay and Monigl (1992); Prague Institute of Transportation Engineering (1994); Pucher, (1993); Rataj (1988) and Suchorzewski (1993).

all four countries between 1980 and 1988. Moreover, there was only a slight decline in public transport's percentage of urban travel (see Table 8.3). In Czechoslovakia, Poland and Hungary, public transport continued to enjoy a very high modal split, with about 80–85 per cent of total motorized travel (excluding walking and bicycle trips).

A number of studies indicate that car use in Socialist countries increased more slowly than car ownership (Grava, 1984; Klofac, 1989). Indeed, even automobile owners usually used public transport for the journey to work because of high operating costs, parking problems and frequent, cheap public transport services during the peak hours. The main use of the car in Socialist Central and Eastern Europe was for social and recreational purposes, such as for weekend trips to the countryside.

Developments in East Germany were not quite so favourable to public transport as in other Socialist countries. Increased automobile use there had begun to erode the dominance of public transport well before the end of Socialism. Overall trip rates, and thus levels of mobility, increased from 2.35 to 2.93 trips per person per day between 1972 and 1987 (Pucher, 1994). Whereas the number of public transport trips per capita increased by only 24 per cent, the number of car trips increased by 85 per cent, almost a doubling of car travel. As a consequence of that unequal growth, public transport's share of motorized journeys (excluding walking and

Table 8.3 Trends in public transport's share of urban travel in Eastern and Central Europe, 1970–92 (% of total motorized trips)

	Poland	*Hungary*	*Czechoslovakia*	*East Germany*
1970	93	92		
1972				62
1975		90		
1977				60
1980	80	88	89	
1982				57
1985		86	85	
1987	79			52
1990		78	80	
1991				35
1992	61	72	79	

Sources: Ministries of Transport of Poland, Hungary, Czechoslovakia, Slovakia, the Czech Republic and Germany; Klofac (1989); Lijewski (1987); Mitric (1994); Polish Association of Urban Public Transport (1994); Polish Central Statistical Office (1982 to 1994); Pasquay and Monigl (1992); Prague Institute of Transportation Engineering (1994); Pucher (1993); Rataj (1988), and Suchorzewski (1993).

cycling) fell from 62 per cent in 1972 to 52 per cent in 1987 (see Table 8.3).

Although public transport declined slightly in relative importance in Eastern Europe during the 1970s and 1980s, it maintained a dominant role in urban travel. That dominance is all the more striking in contrast to the declines in public transport use and modal split in Western Europe. By 1989, for example, public transport's share of urban travel in West Germany was only 11 per cent, less than half the East German level (see Chapter 3). Moreover, between 1980 and 1988, public transport use had fallen even in absolute terms in West Germany (by 15 per cent), whereas it had risen slightly in East Germany (Table 8.2). Urban transport in Czechoslovakia, Poland and Hungary was even more dominated by public transport, with the modal split share of automobile travel only a third as high as in East Germany (10–15 per cent as against 38 per cent).

The dominance of public transport in the Socialist countries of Central and Eastern Europe was strongly supported by public policies that discouraged car use. Socialist governments directly set the costs of ownership and operation very high through their system of regulated prices; in addition, they sharply restricted car production, thus keeping supply limited. In East Germany, for example, the average waiting time between ordering and purchasing a new car exceeded 10 years, and the quality of automobiles was quite low. Moreover, the road network was primitive by Western standards, and there was a severe shortage of petrol stations, repair shops and other service facilities for cars.

In contrast, Socialist governments offered extensive public transport services at extremely low, subsidized fares. The combination of anti-car and pro-public transport policies produced a strong financial incentive to use public transport instead of private cars. In 1988, for example, a litre of petrol cost nine times more than a ride on public transport in East Germany (Pucher, 1990). Likewise, a litre of petrol cost 9–12 times more than public transport fares in Czechoslovakia, Poland and Hungary. In Poland, moreover, petrol was rationed from 1981 to 1988, leading to a black market in ration coupons that further increased the price of petrol for anyone wanting to drive more than was possible with the official allotment of 24–45 litres per month.

The relative price of car use *vis-à-vis* public transport was roughly 10 times higher in Eastern Europe than in Western Europe. For example, a litre of petrol in West Germany cost slightly less than the regular, one-way fare on public transport.

The much higher levels of car ownership and car modal split in East Germany relative to Hungary, Poland and Czechoslovakia were not due to public policies more favourable to the automobile or less favourable to public transport. Rather, the explanation lies in the higher per capita income and standard of living in East Germany, which was the highest of any Socialist country in the world. According to figures of the World Bank, East Germany had a per capita income about 25 per cent higher than Czechoslovakia over the two decades of the 1970s and 1980s, and about 50 per cent higher than Hungary and Poland (Pucher, 1990). Thus, the East German economy was better able to afford higher levels of car use, as were individual German travellers, even before the German monetary union in 1989 and German political unification in 1990.

URBAN TRANSPORT AFTER THE FALL OF SOCIALISM

The economic and political revolutions of 1989 and 1990 dramatically changed the context of transport policies in virtually all the formerly Socialist countries of Central and Eastern Europe. The countries have not yet fully adjusted their policies to reflect the new situation. Until now, most governments have done little to prevent the increase in car ownership and use or to preserve the pre-eminence of public transport. Perhaps they view the continuing shift towards more car use and less public transport as inevitable. To some extent, that is probably true. There is widespread popular demand for more automobiles. As the economies of Eastern Europe grow, so will per capita incomes, and that is bound to stimulate further growth in car ownership. In most countries, moreover, government policies are accelerating the shift to the car by curtailing subsidies to public transport while reducing the relative price of car use. There are significant differences among countries, however. Eastern Germany and Poland, for example, have adopted the most pro-car policies, while Hungary and Czechoslovakia have done more to slow down the decline of public transport.

Urban transport changes in Eastern Germany

The situation in Eastern Germany is surely the most dramatic indicator of things to come in the rest of Central and Eastern Europe. Almost immediately after the overthrow of its Communist dictatorship, car ownership skyrocketed and public transport use plummeted. Car ownership per 1000 inhabitants rose from 237 in 1989 to 296 in 1990, 349 in 1991, and 415 in 1992, representing a 75 per cent increase in only three years. Over the same period, public transport ridership fell by almost 50 per cent (German Ministry of Transport, 1993; Pucher, 1994). Already by 1991, the automobile's share of modal split in urban travel was almost twice as high as that for public transport: 34.3 per cent as against 18.8 per cent. After decades of public transport dominance, the modal splits of public transport and the car were reversed in only a few years.

There are a number of reasons for the surge in automobile ownership and use and for the sudden rejection of public transport. Probably the most important reason was economic. Monetary union with West Germany in 1989 gave East Germans immediate access to large sums of Western currency and to the West German automobile market. In spite of record levels of unemployment, underemployment and forced early retirement (amounting to a third of the labour force), the purchasing power of most Eastern Germans has increased. The German federal government in Bonn has been pumping about DM 150 billion ($100 billion) per year into the Eastern German economy. Much of that enormous sum has gone towards privatizing, modernizing and restructuring industry, and aiding state and local government finances. But it has also included direct transfer payments to individual East Germans in the form of welfare support, child subsidies, pensions, social security, unemployment compensation, and education and training stipends. Clearly, that massive infusion of funds into the East German economy has greatly increased the affordability of automobile travel.

In addition to the sudden availability and affordability of the automobile, the role of the automobile as a symbol of freedom and prosperity has also contributed to automania in Eastern Germany. The car is not simply a mode of transportation; it is an important status symbol. In particular, the very possession of a Western car seems to serve a psychological need for Eastern Germans who are trying to catch up with the much higher standard of living in Western Germany. Förschner (1992) argues that the purchase of a Western German car is one way for Eastern Germans to satisfy their dream of Western German prosperity as soon as possible: it is a sort of 'advance' on the prosperity forecast to come.

Government transportation policies have reinforced these incentives. For example, public transport fares were increased 10-fold between 1990 and 1992. The cost of a regular, one-way ticket rose from only 17 Pfg to almost DM 2 in most Eastern German cities. By contrast, the price of a litre of petrol actually fell, from about DM 1.50 to DM 1.30. Thus, whereas under Socialism a litre of petrol had cost nine times more than a one-way trip on public transport, a litre of petrol now costs less than a public transport trip. This represents a dramatic reversal in the relative prices of the various urban transport modes and obviously discourages public transport use, while making car use more affordable.

Urban transport changes in Czechoslovakia, Poland and Hungary

Transportation developments in Czechoslovakia, Poland and Hungary over the past few years have been similar to those in Eastern Germany. Car ownership and use have increased considerably in all three countries, and the automobile's share of modal split has risen (see Tables 8.1 and 8.3). In Poland and Hungary, public transport use has fallen off considerably, while it has increased somewhat in the Czech Republic (see Table 8.2). There has been a general trend away from public transport and towards the automobile throughout Central and Eastern Europe, but with substantial variations among countries. The modal shift has been enormous in Eastern Germany, large in Poland, moderate in Hungary, slight in Czechoslovakia, and barely perceptible in the countries of the former Soviet Union.

Even under Socialism, car ownership levels in Czechoslovakia, Poland and Hungary fell further and further behind those in Eastern Germany after 1980 (Table 8.1). Moreover, public transport use in Czechoslovakia, Poland and Hungary grew considerably until 1988, whereas it stagnated in East Germany (Table 8.2). Finally, public transport's share of urban travel was much higher in Czechoslovakia, Poland and Hungary than in East Germany (Table 8.3).

All of these differences were greatly magnified upon German reunification. In 1988, East Germany had about 15 per cent more cars per capita than Czechoslovakia, 30 per cent more than Hungary, and 89 per cent more than Poland. By 1992, Eastern Germany had 89 per cent more cars per capita than Czechoslovakia, 91 per cent more than Hungary, and 146 per cent more than Poland. From 1988 to 1992, public transport use plummeted by 56 per cent in Eastern Germany. Ridership also fell in Hungary and Poland, but only by half as much (25 per cent in Hungary, 33 per cent in Poland). In the Czech Republic, public transport use increased by 6 per cent, making it the only country in the region where ridership did not fall. The resulting modal split differences among countries also increased. In 1988, the automobile's share of modal split in Eastern Germany was about twice as high as in Czechoslovakia and Hungary. By 1992, it was three times as high, and the difference is getting larger.

The most important explanation for the increasing differences between Eastern Germany and the other three countries lies in very different trends in per capita income. Whereas Eastern Germans have benefited from a massive infusion of Western German financial aid, the Czech, Polish and Hungarian economies experienced sharp declines in output and personal income and received only minimal financial aid from other countries and international development agencies. Moreover, consumer prices rose by over 100 per cent in both Czechoslovakia and Hungary between 1990 and 1992, while nominal incomes increased by less than 50 per cent, resulting in drastic decreases in real incomes and standards of living in the two countries for the average citizen. Likewise, inflation in Poland far exceeded the increase in average income. Perhaps equally important, Czechs, Poles and Hungarians had only limited access to hard Western currency and thus to Western goods. For all of these reasons, they have not been able to finance a large increase in car ownership, such as in Eastern Germany. The low user cost of public transport has become an even more crucial factor in the choice of travel mode for Czechs, Poles and Hungarians as their real incomes have fallen.

However, there are other reasons as well for differences in recent developments. Public policies in Czechoslovakia and Hungary continue to favour public transport over the automobile to a much greater extent than in Eastern Germany. Public transport fares, for example, remain extremely cheap, both in absolute terms and especially relative to the cost of automobile use. Between 1990 and 1992, regular one-trip fares were increased four-fold in Czechoslovakia and six-fold in Hungary compared to 10-fold in Eastern Germany. Moreover, petrol prices were more than doubled in Czechoslovakia and Hungary, whereas they actually *fell* in Eastern Germany. The ratio of the price of a litre of petrol to the regular, one-way public transport fare fell in Eastern Germany from 9:1 to less than 1:1; the corresponding transport price ratio in Czechoslovakia fell from 9:1 to 5:1, and in Hungary, from 12:1 to 4:1. Thus, the cost of car use relative to public transport fell in all three countries, but much more dramatically in Eastern Germany.

The situation in Poland is somewhat closer to that in Eastern Germany. The so called 'big bang' approach to economic reform adopted in Poland in 1990 dictated drastic reductions in subsidies and large increases in user charges for government services. As a result, public transport fares were raised sharply. For example, between 1988 and 1994, the price for a regular, one-way bus or tram ticket in Warsaw rose 400-fold, from 15 zl to 6000 zl (Mitric, 1993; Mitric and Suchorzewski, 1994). The monthly cost of using public transport rose from 750 zl to 300 000 zl, and the proportion of average monthly income required to pay for the cost of using public transport rose from only 1 per cent of income in 1988 to 6 per cent in 1994. That represents a truly dramatic increase in the price of urban public transport relative to the standard of living.

Equally striking is the change in the price of public transport relative to the price of car use. In 1988, the price of a litre of petrol was eight times as expensive as the price of a one-way bus or tram ticket (120 zl as against 15 zl). By 1994, the price of a litre of petrol was only about twice as high as the price of a public transport trip (10 000 zl as against 6000 zl). Thus, the relative price of car travel has fallen sharply, while the relative price of public transport has risen.

At the same time, petrol rationing came to an end, thus increasing the supply of petrol available for purchase. Moreover, the opening up of Poland to Western markets greatly expanded the number and variety of automobiles Poles had to choose from. Both the quantity and quality of automobiles improved. Just as car travel was becoming more attractive, public transport subsidies were slashed as part of the government's radical economic reform programme. The quality and quantity of public transport services fell. All of these developments have provided a strong incentive to switch from public transport to the automobile.

It is noteworthy that the largest drops in public transport ridership occurred in Eastern Germany and Poland, where the relative price of public transport increased the most, while ridership has increased in the Czech Republic, where the relative price of public transport increased the least. That pattern is also borne out by the 1994 ratios of petrol prices to public transport fares: 0.7:1 in Eastern Germany, 1.4:1 in Poland, 3.1:1 in Hungary, and 3.3:1 in the Czech Republic. In addition, the monthly passes used by most of the population remain deeply discounted in Hungary and the Czech Republic. They are an especially good bargain in the Czech Republic, where monthly ticket prices rose by a much smaller percentage than the one-trip fare. Indeed, over the past four years, the monthly ticket price increased less than the price of petrol and also less than the general rate of inflation, indicating a reduced real price of public transport for monthly ticket holders. By comparison, the price of a monthly pass became much more expensive in Poland, increasing by almost as much as the one-trip ticket price, and by much more than the price of petrol.

Of course, there are other reasons as well for the sharp decline in public transport use in most Eastern European cities. From 1988 to 1992, the unemployment rate rose in Poland from less than 1 per cent to over 16 per cent, and in Hungary, from less than 2 per cent to over 13 per cent (*The Economist*, 1994). That obviously has eliminated many commutation trips formerly made by public transport. The situation has been similar in Eastern Germany, where the unemployment rate rose from less than 1 per cent in 1988 to about 15 per cent in 1994 (Pucher, 1994). Rapid privatization in Poland, Hungary and Eastern Germany accounts for much of the increase in unemployment. By comparison, the unemployment rate in the Czech Republic remains quite low: only 4 per cent in 1994 and 3 per cent in 1995. That low rate of unemployment helps to keep commuter use of public transport high, in contrast to Poland, Hungary and Eastern Germany.

Financial incentives obviously do not provide the only explanation for the modal shift from public transport to the automobile. Throughout Central and Eastern Europe, as well as in the countries of the former Soviet Union, the automobile has become an extremely important symbol of freedom and social status. Interviews conducted by the author with sociologists, urban planners and transportation experts in former Socialist countries confirmed again and again that the demand for automobiles in Eastern Europe far exceeds what is actually necessary to meet mobility needs. The very possession of a Western automobile has become so important for displaying one's social and economic status that it would not be too extreme to speak of automania, a virtual obsession with the automobile. Many Eastern Europeans with quite modest incomes have purchased cars even though they cannot really afford them, because having an automobile is essential to show that one does not belong to the lower class. Public transport use, by comparison, is increasingly associated with lower status and the many decades of Communist repression, when public transport was forced on the population out of Socialist ideology.

Finally, many of the difficulties of public transport in Eastern Europe undoubtedly lie with the public transport systems themselves. Thanks to decades of massive subsidies, the monopolistic structure of the industry, the predominance of political goals and an almost entirely captive ridership, public transport systems became extremely inefficient. As the end of Socialism neared, labour productivity was low, per unit costs were high, overstaffing was extreme, and the quality of service was low (Office of Urban Transport Economics, 1992; Pasquay and Monigl, 1992; Mitric, 1993; Suchorzewski, 1993; Mitric and Suchorzewski, 1994). Moreover, with the exception of generous funding dedicated to building and extending metro systems in Prague, Budapest and Warsaw, the maintenance and improvement of public transport infrastructure and vehicles were woefully neglected, leading to their steady deterioration and ever slower, less reliable service (Pasquay and Monigl, 1992; Mitric, 1993; Ryder, 1994). The fall of Communism brought an abrupt end to massive subsidies for public transport and simultaneously exposed public transport to sharply increased competition from the automobile. Public transport firms are now trying to adjust to the new situation through restructuring, better management, staff reductions and cutbacks in the most unprofitable services, but it will take years before the transformation is complete and, in the meantime, more and more customers are lost forever to the automobile.

URBAN LAND USE PATTERNS

One might suppose that transportation and land use were well coordinated in Eastern Europe and the Soviet Union, since planning played such an important role there. On the contrary, the available studies of urban planning and transportation in those countries indicate surprisingly poor coordination (White, 1979; Grava, 1984; French, 1987; Lijewski, 1987; Klofac, 1989). That was partly the result of inadequate cooperation and even competition among rival ministries individually responsible for industrial location decisions, housing and transport. Equally important was the desperate need for more housing in Eastern Europe under conditions of widespread resource shortages, which forced governments there to adopt expedient short-term solutions to their housing crises that exacerbated their urban transport problems. Because of greater land availability and lower construction costs in the suburbs, almost all new housing over the past few decades was constructed in massive high-rise apartment complexes at the periphery of cities, a policy that has led to extremely long and time-consuming commuter journeys to work (White, 1979; Lijewski, 1987; Klofac, 1989). In addition, the suburban apartment blocks are rarely equipped with adequate shopping, cultural, educational and recreational facilities, so that long trips to the city centre are often necessary for those purposes as well. Although new industrial plants were also located on the periphery, likewise because of the greater availability of land, there was virtually no effort to match up suburban residences with suburban workplaces. That was probably because industry in the Socialist countries of Eastern Europe was extremely polluting, making nearby residential locations highly undesirable. Such functional segregation of the suburbs has resulted in massive cross-commuting. In short, both residential and industrial location decisions made by Socialist planning ministries greatly increased transport requirements and led to excessive travel times. Even the official Communist government documents on social policy cite that as one of the most severe problems afflicting their cities (White, 1979).

The urban spatial structure developed over decades of Socialism will remain a troublesome legacy for many years to come. Even if the countries of Eastern Europe currently had enormous resources at their disposal, the massive, monolithic apartment complexes on the urban fringes would remain where they are for a long time, as would the locations of industrial and commercial enterprises. In fact, there is still a severe shortage of housing in Eastern Europe and a lack of financial resources to deal with a whole range of social, environmental and economic problems. Thus it is certain that the current land use patterns of cities in formerly Socialist countries will persist and will continue to cause severe transport problems. Indeed, as an ever larger proportion of travel is made by automobile instead of public transport, the problems caused by the extreme separation of housing and workplace locations almost certainly will become even worse than previously. The existing roadway network is simply not capable of handling large increases in car travel. As noted in the following section, the result has been alarming increases in roadway congestion, air pollution, noise and traffic accidents.

URBAN TRANSPORT PROBLEMS

The sudden shift from public transport to the automobile in Eastern Europe is causing more severe urban transport problems there than one finds in Western Europe. The land use patterns of formerly Socialist cities are particularly ill-suited to high volumes of automobile travel, and governments in Eastern Europe simply do not have the necessary resources to expand the road infrastructure to handle the increased traffic volumes.

Problems in Eastern Germany

The increase in car use has been most rapid in Eastern Germany, and the resulting problems have also been most serious there. The number of traffic accidents, for example, more than doubled. From 1989 to 1991 traffic fatalities rose from 1784 to 3759 deaths, an increase of 111 per cent in only two years. Similarly, the number of injuries from traffic accidents rose from 41 037 to 83 580, an increase of 104 per cent (Schmidt, 1992; German Ministry of Transport, 1993). The main reason for the appalling increase is the growth in car use. Roadway infrastructure is not sufficient to handle the increased traffic volumes, and many of the existing roads are dangerous, having been built many years ago when road safety engineering was less advanced. Many narrow roads have no shoulders and are often lined by trees directly adjacent to the pavement, without any safety barriers to protect cars straying from the road. Roads in Eastern Germany tend to have more curves and to be more uneven than those in Western Germany; pavements are sometimes so riddled with pot-holes that repaving is imperative. Signage and traffic signalling at intersections is rather outdated, and need to be improved. Moreover, Eastern German drivers are not used to driving the more powerful Western German automobiles, which permit speeds that are unsafe on the primitive Eastern German roads. Driving behaviour, in general, seems to have deteriorated as well, with fast driving viewed as another expression of the new freedom. Speeding and reckless driving have been difficult to deter, as the police system in Eastern Germany broke down after the fall of Socialism, and it has still not recovered (Schmidt, 1992). Crimes of virtually every sort have greatly increased.

Other problems exacerbated by increased car use include congestion, air pollution and energy use. Traffic congestion has for the first time become a serious problem in Eastern German cities, slowing down not just automobiles but also buses and trams and thus diminishing the quality of public transport service as well. The survey by Förschner and Schöppe (1992) found that between 1987 and 1991 the average trip time in Eastern German cities increased by 40 per cent for car travel and by 18 per cent for public transport. Since the average trip length hardly changed at all, those survey results suggest a sharp reduction in average speeds. Voigt's measurements of changes in travel speeds in Dresden confirm that (Voigt, 1991). Speeds on selected roadways were monitored between 1990 and 1991. Over that one-year period alone, average speed fell by 13 per cent during the early morning peak hours, by 18 per cent in the late morning, and by 22 per cent during the afternoon peak. Figures would obviously vary for different cities, but all observers agree that roadway congestion has become much worse since 1989. Moreover, parking facilities are overflowing with more automobiles than they were ever intended to handle, so that the incidence of illegal parking has increased greatly. That blocks roads, thus exacerbating the congestion problem yet further, and also creates additional traffic safety problems.

The overall increase in travel and the modal shift from public transport to the automobile have added to the already serious air pollution levels in Eastern Germany, which are considerably higher than those in most Western German cities. As in virtually all countries of Central and Eastern Europe, Socialist governments almost entirely ignored the adverse environmental impacts of pollution. The closing of most Eastern German industrial plants over the past three years due to their outdated technology and unprofitable operations has greatly reduced industrial pollution levels. That has tended to offset the impact of more automobile pollution.

Finally, the shift to the car has created critical financial problems for public transport systems, which depend on high volumes of passengers to run efficiently. As mentioned previously, systems are being forced to raise fares and curtail services, due to reductions in both passenger fare revenues and subsidies. If this trend continues, it may significantly reduce the mobility of those segments of the population that are either financially or physically unable to drive a car. By restricting access to employment, education, recreation and shopping, a reduction in public transport services would further magnify the ever-growing social inequalities that have arisen with the shift from Socialism to a private market economy.

Problems in Poland, Czechoslovakia and Hungary

As in Eastern Germany, increased car use in Poland, Czechoslovakia (now the Czech Republic and Slovakia) and Hungary has caused some serious problems: increases in traffic accidents, congestion, air pollution and energy use (Institute of Transportation Engineering, 1991; Hungarian Ministry of Transport and Communications, 1992; Suchorzewski, 1993; Mitric and Suchorzewski, 1994). In Hungary, for example, there was a 43 per cent increase in the number of traffic deaths between 1988 and 1990, and a 29 per cent increase in the number of traffic injuries.

Fortunately, the number of traffic fatalities fell by 14 per cent between 1990 and 1992. The explanation for that decline lies in stricter enforcement of traffic regulations in response to the alarming rise in fatalities during the previous two years. Nevertheless, the traffic fatality rate in Hungary remains one of the highest in Europe, at 24 deaths per 100 000 population, compared to an average European rate of 15 traffic deaths per 100 000 (Bruehning, 1993).

In Czechoslovakia the number of traffic fatalities rose by 34 per cent between 1988 and 1990, and the number of serious injuries rose by 26 per cent. Unlike Hungary, however, the number of traffic fatalities in Czechoslovakia continued to increase from 1990 to 1992, indeed by another 19 per cent, resulting in an overall increase in the fatality rate from a very low 9 per 100,000 population in 1988 to 16 in 1992. The number of serious injuries also continued to increase from 1990 to 1992: by another 20 per cent.

Traffic fatalities in Poland have risen even faster than in Czechoslovakia and Hungary, indeed almost as dramatically as in Eastern Germany. By 1991, there were 7901 traffic deaths, an increase of 71 per cent over the 4625 fatalities in 1987. During the same period, the traffic fatality rate per 100 000 population rose from 12 to 21, almost doubling, and reaching a rate nearly as high as in Hungary.

It is particularly disturbing that in all formerly Socialist European countries, a large percentage of traffic fatalities are pedestrians and bicyclists hit by automobiles: 40 per cent in Eastern Germany, 43 per cent in Czechoslovakia, 49 per cent in Poland and 46 per cent in Hungary. In cities, the percentages are even higher. Thus, increased car use has also made walking and cycling much more dangerous. In contrast to car occupants, pedestrians and cyclists are almost totally unprotected against possible injury from collisions. The fatality rate per kilometre travelled is many times higher for pedestrians and cyclists since they make much shorter trips on average than do automobile users.

Eastern Germany, Czechoslovakia, Hungary and Poland all had per capita traffic fatality rates lower than the West European average in 1987, when they were still Socialist countries. Since then, traffic fatalities have risen so sharply that they now have rates higher than the West European average. This is obviously a very serious problem throughout Eastern Europe, and as car ownership and use continue to increase, the problem threatens to become even more severe. As in North America and Western Europe, building safer cars and roads will be essential to reducing car occupant deaths, but it will be at least as important to undertake a whole range of measures to increase the safety of pedestrians and cyclists, since they account for roughly half of all traffic fatalities.

Statistics on other types of transport problems are far less available than those for traffic fatalities. Moreover, they also tend to be less reliable and less comparable. Nevertheless, all indications are that traffic congestion, air pollution and noise from cars, energy use and mobility problems of the poor have all become more serious since the fall of Socialism.

Air, water and ground pollution have been severe problems in Eastern Europe for decades but were largely ignored by Communist governments which single-mindedly pursued the goal of fastest possible industrial growth at all costs. On top of that legacy of environmental deterioration, increased car use has especially exacerbated the problem of air pollution in Polish, Czech and Hungarian cities. Exact statistics on air pollution in Eastern Europe are not available, but various sources indicate that the problem is severe and getting worse (Office of Urban Transport Economics, 1992; Suchorzewski, 1993; Mitric and Suchorzewski, 1994). Similarly, transport noise has become more prevalent. For example, 31 per cent of Warsaw residents are exposed to over 60 decibels, and 5 per cent are exposed to over 70 decibels (Suchorzewski, 1993).

Traffic congestion is probably the most visible problem of increased car use. In Warsaw, for example, a typical rush-hour car trip of 10 km required 20 minutes in 1988; the same trip now requires an average of 30 to 45 minutes, depending on the rather erratic traffic conditions that prevail. The increased travel time implies a reduction in travel speed from 30 km per hour in 1988 to 14–20 km per hour in 1994. Another estimate puts the average daytime travel speed at 18–20 km per hour for Central Warsaw and 23–40 km per hour for parts of Warsaw outside the centre (Mitric, 1993; Suchorzewski, 1993). Bus travel in Warsaw has been especially hurt by the increased traffic congestion because Polish cities have no provisions whatsoever for traffic priority treatment for public transport. The current average speed of bus transport in Warsaw is only 8 km/hour. That slow speed of buses further reduces the perceived quality of service and discourages yet more passengers. The sharp increase in car use and insufficient road capacity in all Eastern European cities have led to worsening traffic congestion in the Czech Republic and Hungary as well. Moreover, as in Poland, urban congestion has particularly slowed down buses, making them even less competitive with the automobile.

Parking problems in East European cities abound. There is a severe shortage of parking places, leading to cars being parked on walkways and streets in ways that block pedestrian, car and bus traffic and cause significant safety problems (Suchorzewski, 1993; Mitric and Suchorzewski, 1994).

Finally, increased car use has drawn away much of public transport's customer base, leading to revenue losses, the need for fare increases and service cutbacks, and yet more ridership loss. For the increasing number of Poles, Czechs and Hungarians with an automobile that may seem an irrelevant problem, but it is causing ever more severe mobility problems for the rapidly growing underclass of the poor.

The slower rate of increase in car use in Czechoslovakia and Hungary has helped to keep the magnitude of car-related transport problems somewhat in check. The more sudden growth in Poland, and especially Eastern Germany, has caused more severe problems. The slower the rate of change, the more time policy-makers, transportation planners and individual travellers have to adjust their policies, plans and travel behaviour to the evolving, more auto-oriented, transport systems. In Eastern Germany, one gets the impression that developments are so rapid that they are simply out of control, and that the rush to ever higher levels of car use ignores its many negative consequences.

CONCLUSIONS AND POLICY IMPLICATIONS

Car ownership and use will certainly continue to increase over the coming years in Central and Eastern Europe. With the overthrow of Communist dictatorships, transport policies in East Germany, Poland, Hungary and the Czech Republic are now determined democratically, which means that the underlying preferences of individual citizens will help to shape those policies. It seems inconceivable that any public policy would now be accepted that restricted car ownership and use as much as under Socialism. For better or worse, the public's demand for car travel seems almost boundless. Its comfort, privacy, flexibility, convenience and speed represent irresistible advantages for most people. Except for academics, environmentalists and urban planners, there does not seem to be much opposition to increased car use in Central and Eastern Europe. Most of the public that continues to use public transport does so not necessarily out of preference but because they cannot afford to buy an automobile. As soon as they can afford an automobile, most Czechs, Poles and Hungarians – just like their German neighbours – will buy and use it extensively, in spite of all its known deleterious impacts on the environment, traffic congestion, safety and energy use. Whether analysts like it or not, the car revolution in Central and Eastern Europe is inevitable.

Even if the move to a car-based transportation system is inevitable, it will be essential to introduce policies such as those in Western Europe to 'tame' the automobile by controlling the nature and extent of car use in order to minimize adverse impacts (Pucher and Clorer, 1992). Examples of such policies include: car-free pedestrian zones; parking restrictions; right-of-way priority for public transport, bicycles and pedestrians; strict enforcement of traffic regulations; minimizing car traffic in central city areas; diverting long-distance traffic around cities altogether; and enforcement of stringent emissions standards.

Technological improvements will be an essential part of any policy to deal with transport problems in Eastern Europe. There is a desperate need to require catalytic converters on cars and to eliminate lead from petrol. Experience from the USA has shown this to be the most cost-effective means of reducing automotive air pollution (Wachs, 1993). Similarly, technological advances in design have been very effective in making cars safer, less noisy, and more energy efficient. Central and Eastern Europe can benefit from over two decades of American experience with such improvements in car design. Nevertheless, technological fixes cannot solve all transport problems.

Perhaps even more important is the need to set the price of car ownership and use at levels that reflect not only direct economic costs but also indirect social, environmental and external economic costs. As described earlier in this chapter, the enormous increase in car use in Eastern Germany has exacted a particularly high toll in traffic deaths and injuries, traffic congestion, noise and air pollution; yet the relative price consumers have had to pay for car use in Eastern Germany has fallen sharply in recent years. That price signal, which strongly encourages more car ownership and use, is precisely the opposite of the message that should be sent to travellers considering their choice of modes. The distortion of the transport price ratio has not been so extreme in Poland, Hungary and the Czech Republic, but even there the relative price of car ownership and use has fallen substantially. If those countries are to avoid the severe transport problems caused by the rapid increase of car use in Eastern Germany, they would be well advised to refrain from further decreases in the relative price of car use.

Unlike the situation under Socialism, travellers should indeed have the choice of automobile use, but they should be required to pay the full social and environmental costs of doing so and not an artificially low price that reflects only direct economic costs. Properly pricing automobile use would probably be the most effective way to limit excessive car use and to encourage support for public transport.

Corrective pricing measures must be coordinated with physical limitations on car use. Parking policy in most Eastern European cities is virtually non-existent, and any parking regulations that do exist are poorly enforced and thus mostly ignored by auto drivers (Mitric, 1993; Suchorzewski, 1993). The consequence is road as well as walkway blockages. Local governments need to restrict the supply of legal parking in central city areas so as to discourage car use on congested routes, and fees for parking should be high enough to reflect at least some portion of the congestion costs arising from car use in densely-developed central city areas. Illegal parking should obviously be discouraged through some combination of towing and fines.

There is also a desperate need to provide some sort of priority treatment for buses on urban streets. The high and growing levels of roadway congestion in Eastern European cities particularly slows down buses, since they have to weave in and out of traffic to pick up passengers at the kerbside. For example, the average speed of bus travel in central Warsaw is only 8 km/hour, which makes it impossible for buses to compete with the automobile. The obvious solution is bus lanes or HOV lanes for buses and car-pools.

Unfortunately, there has been much opposition by car drivers to any sort of limits on their driving (Office of Urban Transport Economics, 1992; Mitric, 1993; Suchorzewski, 1993). That opposition has translated into political blockage of local government measures to control cars use either by restricting lane usage or parking. Of course, the congestion and parking problems may eventually become so severe that even car drivers agree that something must be done. Until now, however, there seems to be an almost allergic reaction to any government policy that even remotely smacks of the behavioural controls so hated during the Communist era. That does not bode well for the future of public transport in Eastern Europe, since doing nothing means a failure to internalize the external costs of car use, which puts public transport at an enormous competitive disadvantage.

The sharp reductions of subsidies to public transport in Eastern Europe are quite understandable from the perspective of local governments, which simply cannot afford to pay out large subsidies any longer. Unfortunately, those reductions put public transport at the mercy of a competitive market which is biased towards the automobile at any rate. Further fare increases, service cutbacks and passenger losses are the inevitable result.

Changes in overall transport policy are crucial to establishing a 'level playing field' for fair competition between the car and public transport. In addition, however, public transport systems themselves can improve their situation by increasing productivity and raising service quality. Many urban public transport systems in Poland, Hungary and the Czech Republic are already in the process of restructuring. The elimination of unnecessary personnel, cutbacks in the most unprofitable services and contracting out part of their managerial, maintenance and operations functions will certainly help reduce costs and thus make public transport more competitive.

Nevertheless, it is hard to imagine improvements in productivity and service quality so dramatic that the decline of public transport in most Eastern European cities will be halted, let alone reversed. Without dramatic government intervention soon, current trends may do irreversible damage to public transport systems. Experience in the USA and Western Europe shows that once public transport riders are lost to the automobile, it is almost impossible to win them back (Pucher, 1988). Until the external costs of car use are fully internalized, subsidies to public transport are essential to prevent a serious distortion in modal choice.

Now is the crucial time to preserve and improve the existing public transport systems in Eastern Europe. Funding for public transport will have to be increased to ensure its continued existence as a viable alternative to automobile use. Unfortunately, neither the central governments nor the local governments in Eastern Europe can currently afford the subsidies needed to improve public transport infrastructure, vehicles and operations. Moreover, outside lending agencies such as the World Bank, the European Bank for Reconstruction and Development and the European Investment Bank have so far neglected public transport and focused instead on roadway improvements (Hook, 1994). Large loans and subsidies are being granted for road projects in Poland as well as Hungary, Bulgaria and Romania. In addition, the same international lending agencies are making substantial investments in improving car and truck manufacturing facilities in Eastern Europe.

There can be no question that investments in roads and motor vehicle manufacturing are needed to deal with the growth in motor vehicle use in Eastern Europe. It is essential to produce safer, less polluting, more fuel-efficient motor vehicles and to improve roads to make them safer and less congested. Nevertheless, the current approach appears to be decidedly one-sided, favouring road and vehicle manufacturing investments while neglecting the desperately needed improvements to public transport systems. Likewise, international agencies have strongly encouraged much higher user charges for public transport while ignoring the serious under-pricing of car use. Such unbalanced public policies will further distort modal choices, exacerbating the already serious social and environmental problems caused by the sudden shift of passenger travel from public transport to the automobile, and of freight traffic from railways to trucks.

Given the widespread popularity of car travel and the predicted economic growth in Eastern Europe, it would be virtually impossible to prevent further increases in ownership and use. Nevertheless, public transport systems should not be allowed to deteriorate to such an extent that they no longer offer a feasible alternative to automotive transport. Reorganisation, rationalization and some reduction of urban public transport systems may be appropriate, but substantial investment in public transport infrastructure is also necessary. The ideal source of funding for such expenditures would be increased car taxes and fees that would help internalize some of the social and environmental costs of car use.

The situation in Central Europe offers a unique opportunity to preserve the best of the old system dominated by public transport while at the same time allowing for an increase in car ownership and use in response to consumer demands. The best model for Central Europe is likely to be Western Europe, where high-quality, extensive public transport systems exist side-by-side with modern, extensive road networks and high levels of car ownership. Indeed, official policy statements from the Czech, Polish and Hungarian governments express such goals, and the transportation policies of Western Germany are being applied to Eastern Germany at any rate (Institute of Transportation Engineering of Prague, 1991; Hungarian Ministry of Transport and Communications, 1992).

Almost regardless of public policies, however, Eastern Germany is likely to continue its rapid pace of automobilization and move away from public transport while Poland, Hungary and the Czech Republic will continue at their slower pace in the same direction, probably with a lag of at least ten years behind Eastern Germany. Economic growth and increases in real per capita income will largely determine the rate of increase in car ownership in virtually all the countries of Central and Eastern Europe.

9 Canada: Bridge between Europe and the United States

Canada is usually overshadowed by its North American neighbour, the USA. Most discussions of transport policy either completely ignore Canada or lump Canada and the USA together in an aggregate treatment of North America. In fact, the transport situation in Canadian cities is quite different from that in the USA, and much of the difference is due to government policies. The contrast between the USA and Canada in urban transport is especially interesting given the many fundamental similarities between the two countries.

Common historical factors include their origins as European colonies, the timing and westward movement of settlement, the availability of vast expanses of land and other natural resources, and the key role of immigration in population growth. Moreover, economic and demographic developments have been similar over the past few decades. Increases in per capita income, population and urbanization have occurred at roughly the same rate in Canada and the USA, due both to geographical proximity and the extreme interrelatedness of the Canadian and American economies.

In spite of such similarities, most Canadians would insist that their country is distinctly different from the USA, especially in culture and lifestyle. For example, Canadian cities have a European ambience lacking in American cities. In many respects, Canada acts as a bridge between Europe and the USA. Most Canadian cities are less dense than European cities but denser than American cities. Canadian car ownership and use are higher than in Europe but lower than in the USA. Public transport use in Canada is almost as high as in Europe and much higher than in the USA. Canada's transport and urban development policies have been closer to those of Europe, with effective regional planning, excellent coordination of public transport services, limited urban highway systems and extensive government restrictions on land use.

However different Canadian cities are from their American neighbours to the south, they have been subject to the same economic and technological forces which have been producing urban decentralization and greater car use throughout most of the world. As discussed next, suburbanization is an important aspect of recent urban development in Canada, even though the nature and extent of Canadian suburbs remain rather different from those in the USA.

TRENDS IN URBAN SPATIAL STRUCTURE

In his analysis of North American cities, Yeates (1990) shows that both Canadian and American cities have decentralized significantly since 1950. For 20 Canadian metropolitan areas, central city population densities (per square mile) fell from an average of 50 000 in 1950 to 20 000 in 1976. For the same period, central city densities fell in 204 US metropolitan areas from an average of 24 000 to 11 000. Newman and Kenworthy (1989) used 1980 data to compare Toronto with the ten largest US metropolitan areas. They found population density (per acre) three times higher in the Toronto metropolitan area (16.2 as against 5.7 in the US cities) and employment density about $2\frac{1}{2}$ times as high (8.1 as against 2.8). Even in the suburbs, densities of both population and employment were about three times higher in Toronto than the average for the 10 largest US cities.

By comparision, Newman and Kenworthy found Toronto less compact and less dense than the 12 largest Western European cities. Overall population density for the 12 European cities averaged 21.9 persons per acre (compared to 16.2 in Toronto). Overall employment density averaged 12.6 jobs per acre (compared to 8.1 in Toronto). European suburbs were also considerably denser than Toronto's suburbs (17.4 and 13.8 inhabitants per acre, and 8.1 and 5.8 jobs per acre, respectively). Thus, the density of Canadian cities lies between the much lower American levels and the higher European levels.

In order to gauge the compactness of urban development, Newman and Kenworthy also calculated the proportions of total metropolitan population and employment in the central city. Similar to the density variations among countries, they found that Canadian cities have a lower percentage of total population and employment in the suburbs than American cities, but a higher percentage than European cities. In short, Canadian cities seem to have land use patterns that bridge the American and European extremes.

There are various reasons why urban development in Canada is more compact than in the USA, but the most important explanation lies in strong government policies that discourage the very low density sprawl found around every American city. Suburban development, for example, is limited to areas that are already served by public water and sewer systems

(Yeates, 1990). Private developers are required to finance in advance any necessary extensions of public utility systems. That has made suburban development much less profitable than it is in the USA. Indeed, by limiting the supply of developable land, Canadian land use policies have increased the price of land, thus encouraging higher density development as a way to save on land costs.

Another public policy that encourages compact urban development in Canada is the explicit coordination of land use and transportation to enhance the viability of public transport as a genuine alternative to the automobile, even in the suburbs. Cervero (1986) and Kenworthy (1991) provide several specific examples of transit-friendly land use planning in Toronto and Vancouver.

The hallmark of government intervention in Canada is the extension of public transport services into the fringes of urban areas prior to residential and commercial development, rather than waiting for development to take place and then responding by providing public transport services afterwards. The necessary rights of way are purchased well in advance, and future public transport services are prominently advertised in order to maximize their impact on location decisions of individual firms and households.

Local governments explicitly plan the establishment of mixed-use suburban centres and ensure that they are well served by public transport services. Far from opposing suburbanization, urban planners in Canada actually promote it in order to relieve the overcrowding of the main centres of metropolitan areas. The type of suburban development they plan, however, is vastly different from the extremely low density sprawl surrounding American cities. It is focused on mixed-use centres, well served by public transport, and much more compact than American suburbs.

Zoning policies in Canada also encourage clustered suburban development, which facilitates public transport use. For example, density bonuses are granted for developments around rail transit stations. In sharp contrast, local governments in American suburbs often down-zone areas around rail stations in an attempt to preserve the low density character of their communities and to avoid the congestion, noise and pollution problems they fear would be caused in the immediate vicinity. While understandable from a local point of view, such actions make it difficult to coordinate regional transportation policy.

Finally, local governments in Canadian urban areas are less fragmented than in the USA, thus facilitating regional planning and making it more difficult for suburbs to 'free ride' on the central city. Canadian cities cannot independently set land use and zoning rules, but must conform to land use guidelines set by provincial regulatory boards. Those provincial guidelines generally favour compact development which can be effectively served by public transport.

TRENDS IN TRAVEL BEHAVIOUR

As already suggested, overall trends in Canadian urban transport are quite similar to those in the USA and Europe. Increased affluence and suburbanization of both population and employment have encouraged ever more automobile travel.

Car ownership and use

Car ownership and use have increased dramatically in Canada over the past four decades. As shown in Table 9.1, the number of cars rose from only 2 million in 1951 to over 13 million in 1991. Some of the increase was obviously due to the doubling in Canada's population during those 40 years, but even on a per capita basis, the car ownership rate more than tripled: from only 139 autos per 1000 population in 1951 to 486 in 1991. That level of ownership is far behind the USA (at about 600 cars per 1000 population), but is higher than almost every European country, and roughly twice as high as the Western European average.

Corresponding to the growth in car ownership, car use has also grown rapidly. Vehicle kilometres of car travel in Canada increased from 1955 km per capita in 1950 to 8230 km in 1990, an increase of 321 per cent

Table 9.1 Trends in population and car ownership, 1951–91

Year	*Population (thousands)*	*Total cars (thousands)*	*Autos per 1000 population*
1951	14 009	1 947	139
1961	18 238	4 085	224
1971	21 568	6 578	305
1981	24 343	10 419	428
1991	27 297	13 266	486

Sources: American Automobile Manufacturers Association (1993); Statistics Canada (1994).

(Statistics Canada, 1993). That rate of increase was twice as fast as in the USA, where car use increased by 155 per cent, from 3838 km per capita in 1950 to 9787 km in 1990 (Federal Highway Administration, 1993).

Public transport use

The long-term trends towards greater car ownership and use in Canada have been reflected in a falling number of public transport trips per capita as well as a declining modal split share for public transport. Although public transport has thus become relatively less important as an urban transport mode in Canada, public transport's decline has been far less severe than in the USA. Canadian cities remained more oriented toward public transport than American cities throughout the period from 1950 to 1990. As shown in Table 9.2, Canada had 246 public transport trips per capita in 1950 (compared to 147 trips per capita in the USA). Four decades later, in 1990, Canada had 104 public transport trips per capita (compared to 38 trips per capita in the USA). Even more striking, public transport's overall percentage of urban travel in Canada in 1990 was roughly five times higher than in the USA (15 per cent compared to 3 per cent), although exact comparisons are not possible due to differences in survey methodology, trip definitions and metropolitan area boundaries.

In terms of public transport use as well as overall modal split shares, Canada lies between the USA and Western Europe. Total passenger-kilometres of public transport use in 1980 averaged 1077 km per inhabitant per year in Toronto compared to only 259 km in the USA and 1288 km in Western Europe (Newman and Kenworthy, 1989). Similarly, the

Table 9.2 Public transport usage trends, 1950–92 (linked passenger trips per year)

Year	*Total trips (millions)*	*Trips per capita*
1950	1 396	246
1960	973	135
1970	980	100
1980	1 315	97
1990	1 519	104
1991	1 450	96
1992	1 402	88

Source: Canadian Urban Transit Association (1980 to 1994, annual).

percentage of total urban passenger-kilometres by public transport was 17 per cent in Toronto, compared to an average of only 4 per cent in the 10 largest US cities and 25 per cent in the 12 largest Western European cities. In short, travel behaviour in Canadian cities lies between the American and European extremes.

Trends over the past few decades have also varied. American public transport systems lost 54 per cent of their passengers from 1950 to 1970, compared to a loss of 30 per cent in Canada (see Table 9.2). Even more significant, Canadian systems not only recovered their lost passengers during the next two decades (1970 to 1990), but even managed to increase usage levels above those in 1950 (from 1.4 to 1.5 billion passenger trips). By contrast, American public transport systems were able to increase usage by only 11 per cent between 1970 and 1990, achieving a 1990 passenger level still less than half of the 1950 level.

One of the reasons for the large increase in public transport use in Canada between 1970 and 1990 is the doubling in the amount of service offered, from 387 million to 781 million vehicle kilometres (American Public Transit Association, 1993). Moreover, during the same period, Canadian public transport systems greatly increased the quality of their services by improving coordination of fares and routes, by constructing new rail lines and busways, by providing extensive park and ride facilities, and by modernizing their information systems.

Since 1990, public transport in Canada has suffered from severe economic recession, with unemployment rates of 10 per cent in 1991 and 11 per cent in 1992. Because work-related trips account for a large proportion of public transport use, reductions in such trips due to unemployment have hit the public transport industry particularly hard. Between 1990 and 1992, usage fell by 8 per cent in Canada, and public transport trips per capita fell from 104 to 88. Compounding the adverse impacts of recession, public transport fares rose by an average of 14 per cent from 1990 to 1992 (6 per cent in excess of inflation), and vehicle-kilometres of service fell by 4 per cent. Thus, the very successful expansion of public transport services during the 1970s and 1980s is now being jeopardized by a period of stagnation and even decline.

The aggregate trends noted above do not reflect the enormous variation in public transport use in different cities. As shown in Table 9.3, there are considerable differences among city size categories, with large cities having much higher proportions of travel by public transport than small cities (for example, 27 per cent in Toronto as against 5 per cent in St Catharines). Similarly, large cities have more public transport trips per capita than small cities (186 in Toronto as against 29 in St Catharines).

Table 9.3 Modal split shares and per capita rates of public transport use in selected Canadian metropolitan areas (1991)

Metropolitan area	*Population (in thousands)*	*Modal split share of public transport*	*Public transport trips per capita*
Toronto	3 893	27	186
Montreal	3 127	34	196
Vancouver	1 603	15	93
Ottawa	921	21	133
Edmonton	840	14	69
Calgary	754	16	68
Winnepeg	652	20	82
Quebec	646	18	80
Hamilton	600	12	60
London	382	10	54
St. Catharines	365	5	29
Halifax	321	n.a.	61
Victoria	288	n.a.	66
Saskatoon	210	n.a.	67
Regina	192	n.a.	40

n.a. = not available.
Note: The population figures used to derive the per capita trip estimates are for the service area, as reported by each transit system, not the total urban area.
Source: Canadian Urban Transit Association (1992).

Even controlling for city size, Canadian public transport accounts for a much higher proportion of urban travel than in the USA. Within the same city size categories, Canadian cities have a public transport modal split three to five times higher than American cities, and they have three to ten times as many public transport trips per capita (Pucher, 1994).

URBAN TRANSPORT PROBLEMS

There is probably less of a crisis in Canadian urban transport than in any other country examined in this book. Of course, virtually every Canadian city has some sort of transport problem, but nowhere in Canada does the situation even approach a crisis. The most serious transport problems appear to be limited to a few of the largest cities and, even for them, the problems are restricted to a few congested corridors during peak hours.

Canada's long-range transport and land use policies have, in fact, been quite successful at preventing serious transport and urban development problems from arising. Moreover, most urban transport problems in Canada seem to be getting less serious over time, reducing yet further the political priority of transport relative to other public policy issues. For example, the number of traffic fatalities has fallen over the past two decades, from 5080 deaths in 1970 to 3501 deaths in 1992. Over the same period, the traffic fatality rate per 10 000 registered motor vehicles fell from 6.0 to 2.0, and the traffic fatality rate per 100 000 population fell from 23.8 to 12.3 (Transport Canada, 1994). Compared to the USA, Canada's fatality rates are lower both per 10 000 vehicles (2.0 compared to 2.1) and per 100 000 population (12.3 compared to 15.4).

Similarly, air quality has been improving in most Canadian cities thanks to reductions in industrial pollution and the use of catalytic converters and unleaded petrol in automobiles. Even the environmental group Greenpeace, which reviewed pollution problems in Canadian cities, found that air pollution is not a significant problem in medium-sized cities such as Winnepeg, Edmonton, Calgary and Ottawa (Vanderwagen, 1992). Automotive air pollution is more of a problem in Toronto and Vancouver but, even there, pollution levels are lower than in most large American cities. Air pollution in Vancouver presents a special problem due to topographical and climatic conditions, which tend to trap pollution, somewhat similar to the situation in Los Angeles. Greenpeace estimates that 75 per cent of Vancouver's air pollution comes from private automobiles. There is a strong environmental movement in Vancouver, and environmental expections are high even among the general population, making transport pollution in Vancouver a high priority issue. For Canada as a whole, however, air pollution is a much less serious problem than in the USA, and the environmental movement in Canada has not had nearly as much impact on transport policy. Indeed, as discussed later, car emissions standards in Canada have been less strict than those in the USA and have simply followed the American lead in pollution control policies anyway.

Energy use in transport is even less serious a problem in Canada, since the country is rich in energy resources relative to its population and thus exports petroleum. Moreover, average fuel efficiency of automobiles has increased dramatically over the past two decades, thanks to technological improvements forced on North American car manufacturers by federal regulations in the USA.

Traffic congestion appears to be most serious in Toronto, Montreal and Vancouver. The topograpical layout of Vancouver around various bodies of water contributes to the problem, since congestion tends to build up at key bridges and tunnels. In Vancouver, Toronto and Montreal, the congestion problem is limited to peak hours and occurs mainly on key suburban arterials and highways leading from the suburbs to the city centre. As in the USA, most of the growth in travel is within and between the suburbs, and that is also where the congestion problem is worsening. In general, however, congestion is not as severe or as pervasive as in American cities, probably due to the better coordination of land use and transport in Canada and the much greater availability of public transport services, even in the suburbs.

The most pressing problem in Canadian urban transport is finance. The Canadian federal government is prohibited by various legislative acts from interfering in urban transport, and that has also meant virtually no federal financial assistance for either roads or public transport. Both provincial and local governments have been struggling with budget deficits and have been forced to cut back their outlays for transport projects as well. That has prevented or at least delayed some crucially needed investments in transport infrastructure as population and travel demands continue to grow. Likewise, fiscal austerity in the public sector has curtailed operating subsidies, requiring fare increases and service cutbacks in urban public transport, which have exacerbated passenger losses caused by the severe economic recession in Canada.

Canadian public transport systems became unprofitable considerably later than their American counterparts, but their increased need for subsidy was also dramatic. As shown in Table 9.4, total operating revenues exceeded total operating costs by 4 per cent in 1970; by 1980, revenues

Table 9.4 Public transport subsidies (in Canadian $), unprofitability and productivity 1970–92

	1970	*1975*	*1980*	*1985*	*1990*	*1992*
Operating subsidy per linked passenger trip (current $)	0	0.15	0.38	0.52	0.74	0.86
Operating subsidy per linked passenger (in 1992 $)	0	0.42	0.73	0.70	0.79	0.86
Operating revenue as % of operating costs*	104	66	54	55	54	53
Vehicle-km of service per employee (in 1000)	17.6	19.4	20.8	20.5	19.7	19.6

*Operating revenue includes passenger fares, advertising and concession revenues and chartered service revenues.
Source: Canadian Urban Transit Association (1980 to 1994, annual).

covered only about half of costs. From 1970 to 1992, the average operating subsidy needed per passenger trip rose from zero to 86 cents, amounting to a total operating subsidy of $1.2 billion on top of a capital subsidy of $451 million. Since 1985, labour productivity has fallen, per unit costs have risen, and the proportion of operating costs covered by passenger revenues has continued to decline in spite of substantial fare increases.

The lack of subsidy funds for improving roadways and public transport systems may have serious long-term consequences. In the past, Canada has been successful at planning well in advance to avoid serious problems. Unless the necessary infrastructure investments are made soon, Canada may find itself in a situation similar to that in the USA, where problems often get neglected, and thus become more and more serious, until they reach a crisis stage when they are more difficult and more costly to remedy.

URBAN TRANSPORT POLICIES

In some respects, transport policies in Canada have been quite similar to those in the USA. For example, Canadians have followed the American lead in requiring car manufacturers to produce more energy-efficient, safer and less polluting cars. Indeed, since many automobiles manufactured in Canada are for the American market anyway, it was virtually inevitable that Canadian standards would be roughly the same as in the USA. The main difference is that Canadian standards are somewhat lower and tend to lag behind the American standards. They also tend to be voluntary agreements negotiated between companies and the federal government, whereas in the USA they are specific standards mandated by federal law, with fines for non-compliance.

Responsibility by government level

Unlike the USA and most European countries, the federal government in Canada has never played an important role in urban transport and is even prohibited by law from doing so (Soberman, 1983). The provinces and municipalities of Canada share in determining urban transport policy, and they are solely responsible for funding operating and capital subsidies.

Although Canadian cities have a considerable amount of autonomy in transport planning, they are not free to do whatever they want to do. In particular, they are dependent to varying extents on transport funding provided by the provincial level of government, which in general is much more powerful than the state level of government in the USA.

As shown in Table 9.5, the provinces have provided a steadily declining percentage of total public transport subsidies since 1980. The provincial share of total operating subsidy funds fell from 53 per cent in 1980 to 42 per cent in 1992, and the provincial share of capital subsidies fell from 89 per cent to 56 per cent. That sharp decline in provincial aid, together with the total lack of federal funding, has caused ever more serious financing problems for Canadian public transport. In particular, it has led to delays in vehicle replacement (and thus an overaged vehicle fleet), and postponement or even cancellation of major infrastructure improvements (such as extending existing rail systems or building new ones).

The situation differs greatly, however, from one province to another. Canadian provinces vary considerably in their transport policies and, in particular, in their funding for urban public transport. As shown in Table 9.6, some provinces are more generous than others. New Brunswick, Newfoundland, Quebec and Saskatchewan, for example, provide no operating assistance at all, whereas British Columbia, Yukon and Nova Scotia finance over half of the total operating subsidy. Similarly, New

Table 9.5 Public transport funding, 1970–92 (Canadian $ millions)

	1980	*1985*	*1990*	*1991*	*1992*
Operating subsidy as % of operating costs	46	45	46	47	47
Total operating subsidy (current $)	502	748	1 134	1 118	1 203
Total operating subsidy (1992 $)	955	997	1 212	1 129	1 203
Provincial share (%)	53	55	52	51	42
Local share (%)	47	45	46	46	55
Total capital subsidy (current $)	196	210	504	486	451
Total capital subsidy (1992 $)	373	280	538	491	451
Provincial share (%)	89	75	67	61	56
Local share (%)	10	23	23	28	35

Source: Canadian Urban Transit Association, *Summary of Canadian Transit Statistics* (1980–1994).

Brunswick, Newfoundland and Saskatchewan offer no capital assistance, whereas Alberta, Ontario and Quebec fund roughly three-quarters of capital subsidies. Not only does the amount of provincial aid vary, but provinces have quite different procedures and formulae for distributing subsidy funds. Table 9.6 also shows that the overall unprofitability of public transport varies from one province to another. Operating revenues cover less than half of operating costs in Quebec, for example, while they cover almost two-thirds of operating costs in Ontario. Of course, the degree of cost recovery from the farebox also varies from one system to another within any given province.

Federal funding is also lacking for urban road construction and maintenance, so that provincial and municipal governments must share road costs as well. The proportion of provincial assistance varies from province to

Table 9.6 Variations by province in government funding of public transport subsidies, 1992

	*Operating ratio**	*Operating subsidy*		*Capital subsidy*	
		% Province	*% Local*	*% Province*	*% Local*
Alberta	0.47	**	**	75	25
British Columbia	0.51	70	30	60	40
Manitoba	0.57	50	50	50	50
New Brunswick	0.42	0	100	0	100
Newfoundland	0.48	0	100	0	100
Nova Scotia	0.61	75	25	50	50
Ontario	0.63	50***	50***	75	25
Quebec	0.47	0	100	75	25
Saskatchewan	0.57	0	100	0	100
Yukon	0.28	100	0	100	0

*Operating ratio is the proportion of total operating costs covered by passenger fare revenues.
**Cities have the option of using a part of their municipal services block grant from the Province of Alberta for public transport, but there is no categorical assistance earmarked exclusively for public transport operating subsidy.
***Actual subsidy received from the Province of Ontario is based on target revenue/cost ratios, which may vary by size of city.
Source: Canadian Urban Transit Association, *Summary of Canadian Transit Statistics*, 1980–1994; Canadian Urban Transit Association (1993).

province, but is generally highest for the most important arterial highways that provide connections to other parts of the province and the rest of the country (Soberman, 1983; Kitchen, 1990; Hutchison, 1991).

Policies on public transport and roadway supply

One of the most important differences between Canadian and American urban transport policy has been in the supply of limited-access highways. The lack of any federal funding for urban expressway construction largely explains the modest extent of highways in Canadian cities. Canada was thus spared the large-scale destruction of central city neighbourhoods that accompanied the massive urban expressway projects in American cities during the 1960s and 1970s. The extremely generous 90 per cent federal funding of urban expressways in the USA made highway construction projects almost irresistible from a local perspective, if for no other reason than to generate jobs for the local economy. Although provincial governments in Canada offer financial assistance to municipalities for highway construction, it is not nearly as generous as federal subsidies in the USA. Because there are no federal matching funds at all in Canada, provincial governments have no exogenous financial incentive to favour highways over public transport. The modest extent of highways in Canadian cities has helped to preserve the attractiveness of central city living and to promote compact, clustered suburban developments that can be served by public transport.

Similar to European cities – and in sharp contrast to American cities – Canadian central cities have not been deserted by the affluent. Although an increasing portion of the middle class lives and works in the suburbs, Canadian central cities remain vibrant, economically viable and pleasant places to live. They are not relegated to being a residence of last resort, inhabited mainly by the poor and working classes, as in many American cities. That has important consequences for transport. As in Europe, public transport serves a broad socioeconomic spectrum of the Canadian population, and thus can draw on a larger potential market.

At the same time as urban highway construction was restrained, local and provincial governments invested large sums in expanding and improving urban public transport systems, especially between 1975 and 1985. As a result, Canada boasts extensive, high-quality metro systems in Montreal and Toronto; modern light rail systems in Calgary, Edmonton, Toronto and Vancouver; highly effective busways in Ottawa; and suburban rail networks in Toronto, Montreal and Vancouver. Recent fiscal austerity in Canada's public sector has been slowing down further expansion of public transport services, but even in the current financially strained environment, Toronto has just approved plans for a major expansion of its metro system.

By restricting highway construction and increasing the amount and quality of public transport over the past two decades, Canada's provinces and municipalities have helped pilot Canadian urban transport down a quite different road from that taken in the USA.

Parking policies

The availability and price of car parking is an important factor in modal choice. As in most European cities, parking is much more restricted in Canadian cities than in the USA. The study by Newman and Kenworthy (1989) reports, for example, that the number of parking places in Toronto per 1000 central city workers in 1980 was 198, only about half as many as the average for the 10 largest American cities (380), and even less than the average for the 12 largest European cities (211).

Part of the reason for the lesser availability of parking in Canadian cities is deliberate policy aimed at reducing car use in the most congested central city areas. Parking is even restricted in suburban centres. For example, Cervero (1986) cites local government building codes for one of Toronto's suburban town centres which set a maximum of 0.3 parking spaces per 1000 square feet of office space. He compares that to the norm of 4.0 parking spaces per 1000 square feet of office space in most surburban office complexes in the USA.

Taxation policies

Canadian taxes on petrol are not nearly as high as those in Europe, but they are more than twice as high as in the USA. According to the OECD (1993), the tax per litre of petrol in Canada was $0.205 (in US dollars) in 1993, compared to only $0.087 in the USA. The average retail price per litre of petrol in Canada was $0.425 (in US dollars) in 1993, compared to $0.301 in the USA.

The higher price of petrol in Canada since 1985 has helped offset the impact of increased public transport fares. By comparison, petrol prices have plummeted in the USA, thus compounding the harmful impact of fare increases and encouraging even more car use. Adjusted for inflation, the real price of petrol fell by 48 per cent in the USA between 1980 and 1993, whereas it rose by 9 per cent in Canada during the same period.

In sharp contrast to the USA, neither mortgage interest nor local property taxes can be deducted from federal income taxes in Canada

(Soberman, 1983). Thus, there has been much less financial incentive for low density residential suburbanization. Perhaps this is one explanation for the lesser degree of suburbanization in Canada and the higher density of the suburbs that do exist. The more compact urban development in Canadian cities has made it easier for public transport systems to reach most of the urban population.

Land use policies

As already mentioned in the discussion of urban development trends, local and provincial governments in Canada have placed much more stringent controls on land use than their counterparts in the USA. Not only has the amount of land available for urban development been restricted, but the price of private land development on the fringes of urban areas has been raised by forcing private developers to pay in advance the costs of necessary infrastructure investment.

The compact development in Canadian cities and suburbs has obviously enhanced the economic viability of public transport, which relies on high employment and residential densities and focussed travel patterns.

CONCLUSIONS

Canadian cities have been quite successful in avoiding the crisis situations facing many cities in the USA. Canada's progressive land use and transport policies have produced cities and urban transport systems distinctly different from those in the USA. Most strikingly, the overall proportion of journeys by public transport has remained high in Canada, almost as high as in European cities. In contrast to the car-dependent cities of the USA, Canadian cities offer a viable transport alternative to the automobile. Similarly, central cities in Canada are much healthier than those in the USA, with larger proportions of urban area employment and population, higher income residents, less crime, better public services, less pollution, less congestion and more vibrant neighbourhoods. In many respects, Canadian cities are more similar to European cities across the ocean than to their American neighbours on the same continent.

It was not inevitable that Canadian cities would develop as they did or that they would be able to avoid the transport crisis facing cities in so many other countries. Forward-looking land use planning was used to coordinate transport with urban development. Large investments were made in extending and improving public transport systems, while at the same time limiting the expansion of urban highways. Higher petrol taxes, limited parking supply, the absence of federal subsidies for home owners and zoning for compact urban and suburban development have all contributed to the much greater success of public transport in Canada *vis-à-vis* the USA.

The current economic recession and fiscal austerity in the public sector has been slowing down further improvement in Canadian public transport systems. Nevertheless, there appears to be widespread commitment in Canada to future investments in public transport when funds become available. Moreover, the range of public policies noted above still discourage urban sprawl and excessive car use. Thus, it seems likely that the success story of urban transport in Canada will continue.

10 The United States: The Car-Dependent Society

No other country in the world is as dominated by the automobile as the USA. From the very beginnings of automobile travel in the early twentieth century, rates of automobile ownership and use in the USA have exceeded levels in other countries, and current rates of ownership and use are by far the highest in the world. Even countries with higher per capita incomes have fewer cars per capita than the USA.

The automobile has not only dominated passenger transport in the USA; it has also become the most important determinant of the American lifestyle, urban form, and even the organization of the American economy. Virtually every aspect of life in the USA – work, social activities, recreation, education and culture – is crucially dependent on the automobile. For most Americans, every other mode of urban transport is practically irrelevant, and life without the automobile is unimaginable. Unlike other advanced industrialized countries, where car ownership only became widespread over the past two or three decades, almost all Americans living today grew up in an automobile dominated society, and most of them have never experienced anything else.

The dominance of the car in the USA is especially striking in cities because its impact on urban land use patterns is highly visible and unmistakable. It is also what most clearly distinguishes American cities from European cities. The term 'urban sprawl' first emerged in the USA to describe the extremely low density, unplanned, rather haphazard residential and commercial development that increasingly surrounds every American city. Widespread suburbanization began earlier in the USA and has been more extensive and lower density than virtually anywhere else in the world. Low density urban sprawl would be impossible without the automobile. Just as the automobile encouraged suburbanization, so suburbanization has encouraged ever more automobile use, since low density development cannot be served effectively by public transport.

The extremely high levels of car use in American cities have caused severe problems of congestion, air pollution, noise, loss of open space, traffic accidents and inadequate mobility for the poor, the elderly and the handicapped. Similar problems have arisen in other countries, but they generally arose earlier in the USA and have been more severe than in other countries. Perhaps for that reason, public policies to deal with the problems of car use were often developed first in the USA. Other countries can learn much from the USA by observing some of the terrible problems of an extremely car-based transport system, and by maintaining more balance in their own transport systems. Moreover, the USA has been successful in developing some effective measures to deal with the social and environmental problems of the automobile, and other countries might benefit from that experience as well. Since most of the Western world seems to be following the American lead in car transport and suburbanization, it is essential to examine the consequences of those developments in the USA.

TRENDS IN URBAN SPATIAL STRUCTURE

American cities have been decentralizing for more than a hundred years (Warner, 1972; Schaeffer and Sclar, 1980; Yeates, 1990). At first, suburban developments were exclusively residential. They emerged along extensions of various types of public transport systems: horsedrawn trams, electric trams, ferry lines, steam railway lines, underground and elevated railways. The type of suburbanization engendered by such public transport modes was rather dense and compact, and was within walking distance of public transport services. The automobile, by contrast, enabled suburban development at very low densities, often not even contiguous to existing development (so called 'leapfrogging'). At the same time as the automobile permitted low density residential suburbs, the motorized truck replaced the horse and wagon for goods transport, thus permitting the suburbanization of firms for the first time.

Since the 1920s, the history of the American city has been one of continuing decentralization, with occasional slow-downs during recessions and wars, but never a reversal in the basic outward trend. Without exception, the average density of urban development has fallen decade after decade, and the trend towards suburbanization of both population and employment is at least as strong today as it was two or three decades ago. From 1950 to 1990, the proportion of the total US population living in metropolitan areas rose from 56 per cent to 77 per cent, but the proportion of the metropolitan population living in the central city fell from 59 per cent to 41 per cent (US Department of Commerce, 1993). The central city's share of total population in 1990 was especially low for large urban areas (as low as 25 per cent in some). The central city's share of total employment was somewhat higher, but considerably less than 50 per cent in most urban areas. Almost all of the growth is now in the suburbs, with most central cities losing both population and jobs.

In recent decades, suburbanization has been occurring in European cities as well, but it has not been nearly as extensive as in the USA. Newman and Kenworthy (1989) compared the forms and densities of the 10 largest American cities and the 12 largest Western European cities. They found that, on average, European cities had a higher proportion of total urban area employment in their central business districts (20 per cent compared to 12 per cent in the USA), a higher proportion of total urban area employment in their central cities (60 per cent compared to 36 per cent), and a higher proportion of their total urban area population in the central city (42 per cent compared to 26 per cent in the USA).

Not only is there more suburbanization in the USA, but American suburbs are much less dense than European suburbs. Newman and Kenworthy found population densities in European suburbs four times as high as in American suburbs, and employment densities were more than three times as high in European suburbs. Central city population and employment densities, by comparison, were only about twice as high in Europe as in the USA. In short, American suburbs are very different from their European counterparts.

American cities are thus much more decentralized than European cities, and they continue to decentralize further at a rapid pace. Most Americans live, work, go to school, shop and find recreational opportunities in the suburbs. They have little reason to venture into the central city at all, except for occasional cultural stimulation or special events. Massive new edge cities have been developing at the fringes of most American urban areas, almost always along major highways, and especially at intersections of major highways. The edge cities are quite segregated in function, with huge office complexes at one site, mammoth shopping centres at another, industrial parks at another, warehousing complexes at another. The failure to plan for mixed-use development or any sort of coordination has led to massive cross-commuting within and among the suburbs. As a consequence, the fastest growing type of urban travel in the USA is the suburb-to-suburb trip, while the historically important suburb-to-central city commute has become decreasingly important (Cervero, 1986 and 1989; Downs, 1992).

Of course, not all American cities are the same. The older cities of the Northeastern USA, such as New York, Philadelphia and Boston, are both denser and less car-dependent than younger cities in the West, such as Los Angeles, Phoenix and Denver. The differences, however, are largely limited to the central cities of those urban areas. Suburban development is uniformly low density in virtually all American cities. Suburban population densities in Chicago, Washington and Boston, for example, were at or below the national average (11, 11 and 10 persons per acre, respectively, compared to an average of 11 for the 10 largest metropolitan areas: see Newman and Kenworthy, 1989). Similarly, suburban employment densities in Chicago, Washington and Boston were also at or below the national average (5, 6 and 4 jobs per acre, respectively, compared to an average of 5 for the 10 largest metropolitan areas). New York's suburban population and employment densities were only slightly above the national average (13 persons per acre and 6 jobs per acre, respectively). Since almost all population and employment growth is occurring in the suburbs, metropolitan areas in the Northeast are becoming ever more like their much younger Southern and Western counterparts. Suburbs in the USA look the same in every part of the country.

These trends in urban development have important consequences for transportation. As American cities become ever more suburbanized, polycentric and low density, travel patterns become ever more diffuse and less focused on any particular centre. In the nineteenth and early twentieth centuries, trips were mostly focused on the one dominant core, the central business district, and that led to high travel volumes in radially-oriented corridors that could be well served by public transport. The vast majority of urban trips today are outside those traditional corridors. They originate in a multitude of different places and terminate in a multitude of different places, generating enough volume to cause congested roads, but not enough in any given corridor to make public transport economically feasible. As urban land use patterns continue to evolve in this direction, American cities are doomed to become ever more car-dependent. That is precisely what has happened over the past few decades, as documented in the following section.

TRENDS IN URBAN TRANSPORT

Consistent travel survey data for the USA do not go back very far, but the available statistics confirm the increasing dominance of the automobile in urban transport. The US decennial Census, for example, has collected commuting information since 1960. Those data indicate that the percentage of worktrips made by automobile has increased from 75 per cent in 1960 to 91 per cent in 1990. Over the same period, the proportion of commuter journeys made by public transport fell from 14 per cent to 5 per cent, and the proportion made by walking fell from 11 per cent to 4 per cent (Federal Highway Administration, 1993).

The Nationwide Personal Transportation Survey has been conducted by the US Department of Transportation since 1969 and shows roughly the same trend, but includes not simply commuter trips but all trip purposes, and thus is far more comprehensive. As shown in Table 10.1, the car's share of all urban travel rose from 80 per cent in 1969 to 84 per cent in 1990. Public transport's modal split share fell from 5 per cent to 3 per cent, and the share of walking trips fell from 12 per cent to 9 per cent (Federal Highway Administration, 1992). It is noteworthy that commuting's share of total travel also fell during this period, from 32 per cent to 26 per cent of all trips. Since public transport accounts for a higher proportion of work journeys than social, recreational or shopping trips, the reduced importance of work trips relative to non-work trips may also help explain the declining modal split of public transport overall. Nevertheless, as already noted, public transport's share of commuter trips has also fallen sharply.

Although most American cities are car-oriented, there is significant variation from one city to another in the degree of car domination. Public transport's modal split share ranges from a high of 28 per cent of all commuter trips in the New York metropolitan area to a low of 2 per cent of commuter trips in the Phoenix metropolitan area (see Table 10.2). Single-occupant auto use ranges from 83 per cent of commuter trips in Detroit to 52 per cent in New York. Car-pooling ranges from 16 per cent in

Table 10.1 Modal split trends in urban travel, 1969–90 (as % of total trips, all trip purposes)

Transport mode	*1969*	*1977*	*1990*
Car	79.8	82.3	84.3
Public transport	4.9	3.4	2.8
Walk	11.5	10.7	9.1
Bicycle	0.7	0.7	0.7
Other*	3.1	2.9	3.1

*This category includes school buses, motorcycles, mopeds and taxis.
Source: *Nationwide Personal Transportation Survey*. US Department of Transportation, Federal Highway Administration (Washington, DC: 1973, 1981 and 1993).

Washington, DC, to 10 per cent in Detroit, New York, Boston and Cleveland. Walking ranges from nearly 7 per cent in New York to less than 2 per cent in Dallas and Atlanta. In spite of this considerable variation, Table 10.2 shows quite clearly that *all* American metropolitan areas have at least twice as much car use as public transport use. Even in New York, which has by far the most public transport service in the country and the densest land use pattern, public transport accounts for less than a third of all commuter trips, and an even lower percentage of non-commuter trips.

One explanation for the continuing modal shift away from public transport and walking is the seemingly endless growth in car ownership. As shown in Table 10.3, total car registrations in the USA almost tripled

Table 10.2 Modal split distributions for work trips in the 20 largest metropolitan areas, 1990 (percentage of all work trips by mode)

Metropolitan areas	*Automobile*		*Public transport*	*Walk*	*Bicycle*
	Solo	*Car-pool*			
New York City	52.3	10.3	27.8	6.5	0.2
Los Angeles	72.9	15.5	4.6	2.9	0.7
Chicago	67.5	12.0	13.7	4.0	0.2
San Francisco	68.8	13.0	9.3	3.6	1.1
Philadelphia	69.2	12.2	10.2	5.3	0.3
Detroit	82.7	10.1	2.4	2.4	0.2
Boston	70.2	10.3	10.6	5.5	0.4
Washington	63.1	15.8	13.7	3.9	0.3
Dallas	78.9	13.8	2.4	1.9	0.1
Houston	76.3	14.6	3.8	2.3	0.3
Miami	75.5	14.5	4.4	2.3	0.6
Atlanta	78.1	12.7	4.7	1.5	0.1
Cleveland	79.6	10.3	4.6	3.0	0.1
Seattle	73.8	11.9	6.3	3.5	0.5
San Diego	71.6	13.8	3.3	4.5	0.9
Minneapolis	76.1	11.2	5.3	3.2	0.4
St Louis	79.8	12.1	3.0	2.2	0.1
Baltimore	71.0	14.2	7.7	4.1	0.2
Pittsburgh	71.5	12.8	8.0	5.1	0.1
Phoenix	75.8	14.4	2.1	2.7	1.4

Note: Columns do not add to to 100% because taxi, motorcycle and various other minor modes are not listed in the table.
Source: Federal Highway Administration (1993).

between 1960 and 1990, increasing from 55 million to 152 million. Some of that growth was due to population increase but, even on a per capita basis, car ownership doubled: from 0.3 cars per capita in 1960 to 0.6 cars per capita in 1990. During the same period, the percentage of households without a car fell from 22 per cent to 12 per cent, and the percentage of households with two or more cars rose from 19 per cent to 37 per cent.

Table 10.4 shows in detail the trends in ridership for each type of public transport from 1950 to 1992. Overall, public transport lost 57 per cent of

Table 10.3 Trends in population and car ownership, 1960–90

	1960	*1970*	*1980*	*1990*
Population (millions)	179.3	203.2	226.5	248.7
Cars (millions)	54.8	79.0	129.7	152.4
Cars per 1000 population	306	389	573	613
Households without car (%)	21.5	17.5	12.9	11.5
Households with two or more cars (%)	19.0	29.3	34.0	37.4
Households with three or more cars (%)	2.5	5.5	17.5	17.3

Source: Federal Highway Administration (1993).

Table 10.4 Trends in urban public transport ridership, 1950–92 (millions of passenger trips)

	1950	*1960*	*1970*	*1980*	*1990*	*1992*
Light rail	2 790	335	172	81	133	144
Heavy rail	2 213	1 670	1 574	1 420	1 455	1 368
Commuter rail	350	310	240	280	328	314
Trolley bus	1 261	447	128	71	106	107
Motor bus	7 681	5 069	4 058	4 774	4 712	4 586
Total	14 295	7 831	6 172	6 626	6 734	6 519

Note: Light rail refers primarily to tramways; heavy rail refers to metros and various types of elevated and underground rail systems within cities; commuter rail refers to suburban trains; trolley bus refers to rubber-tyred buses with electric motors, drawing power from overhead wires; and motor buses refers to diesel-powered, rubber-tyred buses.
Sources: American Public Transit Association (1973 to 1993); US Department of Transportation (1985 to 1994).

its ridership in the two decades from 1950 to 1970. From 1970 to 1992, ridership stabilized, increasing by 6 per cent. Every public transport mode lost ridership from 1950 to 1970, but the losses were particular dramatic for light rail (–94 per cent) and electric trolley bus (–90 per cent). It is noteworthy that the decline in ridership has continued for heavy rail systems, in spite of the new metro systems that were built over the past few decades in San Francisco, Washington, Baltimore and Atlanta. The large losses in the older subway systems in New York, Philadephia, Boston and Chicago have overwhelmed the passenger growth generated by the new systems. Although passenger gains overall have been quite modest since 1970, it was a considerable accomplishment halting the free fall decline of public transport prior to 1970.

Table 10.5 compares various indices of supply and demand for public transport and automobile use. Perhaps somewhat surprisingly, the supply of public transport services expanded considerably more than the supply of urban roads from 1970 to 1992 (58 per cent as against 40 per cent). In contrast, the demand for car travel in urban areas more than doubled (+138 per cent), while public transport ridership increased only slightly (+6 per cent). Thus, the ratio of demand to supply increased dramatically for urban roads, while it actually fell for public transport. The visible consequences of those trends have been ever more congested urban roads and public transport vehicles with fewer and fewer passengers.

One of the reasons why public transport services could be expanded so much since 1970 is that government subsidies to public transport skyrocketed after 1970, increasing roughly 30-fold, from only $0.5 billion in 1970 to over $15.1 billion in 1992.

Table 10.5 Trends in supply and demand for urban roads versus public transport, 1970–92 (all amounts in millions)

	1970	*1980*	*1990*	*1992*
Length of urban road (metres)	898	997	1 211	1 260
Vehicle-km of public transport service	3 013	3 635	4 669	4 755
Car km driven in urban areas	912	1 365	2 040	2 174
Car km per km of road	1.016	1.369	1.685	1.725
Public transport passenger trips	6 172	6 626	6 734	6 519

Sources: Federal Highway Administration (1970 to 1993); American Public Transit Association (1975 to 1993).

By comparison, the net government subsidy to road users (that is, total government expenditures minus road user charges and taxes) increased only three-fold, from $10.6 billion to $32.5 billion (see Table 10.6). Of course, the total government subsidy to roads is still much larger than the total public transport subsidy (about twice as large). Nevertheless, it is astounding that public transport receives 50 per cent of the subsidy given to roads but accounts for only 4 per cent as many trips as car travel. In this respect, the explicit government subsidy per passenger trip is clearly many times larger for public transport than it is for car users. As discussed later, there are a myriad of other, much larger, indirect and implicit subsidies to car users that far outweigh any direct subsidies to public transport. Direct subsidies, however, have been in favour of public transport, at least recently.

URBAN TRANSPORT PROBLEMS

The automobile is the cause of serious social, environmental and economic problems in the USA, just as in other industrialized countries. The long history of automobile dominance in the USA corresponds to a similarly long history of urban problems associated with automobile use. Moreover, the extremely high current level of car ownership and use in the USA, the highest in the world, leads one to expect the world's most severe transport problems in American cities. That is only partly true. For some problems, the long history of automobile use has forced Americans to come to grips with the negative side effects of car use earlier than

Table 10.6 Trends in government subsidy support for roads versus public transport, 1970–92 (all amounts in $ millions)

	1970	*1980*	*1990*	*1992*
Road user charges, taxes and fees	10 248	17 177	44 172	51 848
Government expenditure on roads	20 835	41 795	74 885	84 341
Net subsidy to roads	10 587	24 618	30 713	32 493
Government subsidy to urban public transport	518	7 139	13 371	15 051

Sources: Federal Highway Administration (1970 to 1993); and American Public Transit Association (1975 to 1993).

elsewhere in the world, where widespread car travel is more recent. Moreover, in contrast to Europe, many American cities grew up with the automobile and thus have adapted their land use patterns in ways that mitigate at least some of the problematic aspects. Nevertheless, the automobile has unquestionably been the source of enormous social and environmental problems, especially in urban areas. The extraordinarily high level of car use, together with urban sprawl, have also been partly responsible for the other, largely financial, problems in urban transport. Subsidies to both roads and public transport have increased dramatically. Subsidies to roads have grown in response to the large increase in car travel. Subsidies to public transport have grown partly as an offset to the massive indirect and implicit subsidies to car use and suburban living, which now exceed $500 billion per year. We deal in turn with these two sets of transport problems: first the social and environmental impacts of car use, and then the financial problems of subsidizing urban transport.

Social and environmental problems of car use

Although the automobile obviously generates many benefits to users by permitting unmatched levels of mobility, comfort, convenience and flexibility, it also causes severe social and environmental problems, most of which are not borne directly by car drivers and thus are designated as external costs by economists.

Air pollution

Automobile and truck use account for most of the CO and HC pollution in most American cities, and for a substantial portion of NOx and particulate pollution as well. Prior to the introduction of unleaded petrol, highway traffic also was the major source of airborne lead emissions. The exact percentages vary from one type of pollution to another and from one city to another. Among the 20 largest metropolitan areas, highway traffic's proportion of total NOx pollution ranged from 20 per cent in St Louis to 59 per cent in San Diego, with an average of 39 per cent. Highway traffic's proportion of total HC pollution ranged from 17 per cent in Houston to 74 per cent in Washington, DC, with an average of 59 per cent. Highway traffic's proportion of CO pollution ranged from 28 per cent in Chicago to 95 per cent in Baltimore and Boston, with an average of 77 per cent (Altshuler *et al.*, 1979).

Air quality in America's cities continues to be a serious concern throughout the country, but tremendous progress has been made since

1970, both in reducing automobile emissions and in improving overall air quality (Altshuler *et al.*, 1979; Environmental Protection Agency, 1993; Kessler and Schroeer, 1993; Wachs, 1993). Since 1970, pollution emissions of new cars per vehicle kilometre (so-called tailpipe emissions) were reduced by 91 per cent for HC, by 96 per cent for CO, and by 85 per cent for NOx. Moreover, in spite of large increases in total vehicle kilometres of highway travel, total highway emissions have fallen substantially. In the most recent period, from 1981 to 1992, total highway emissions fell by 46 per cent for HC (volatile organic compounds), by 45 per cent for CO, and by 25 per cent for NOx. Airborne lead emissions have been eliminated almost completely. As a result of emissions reductions, the concentrations of these various types of pollution in the air have been greatly reduced as well, with dramatic reductions in lead, CO and VOC (ozone) pollution, and modest reductions in NOx and particulate pollution (Environmental Protection Agency, 1993; Kessler and Schroeer, 1993). Even Los Angeles, America's most polluted city, has experienced truly dramatic improvements in its air quality over the past two decades, in spite of its continuing bad reputation (Wachs, 1993).

As discussed later in the policy section, all of this progress in reducing air pollution has resulted from technological advances in car and truck design that have reduced the amount of pollution per kilometre travelled. Those reductions have more than offset the impact of the large increase in vehicle km travelled. It is questionable, however, whether future progress will continue at a rate that will offset the impact of future increases in travel.

Congestion

The problem of traffic congestion has become one of the two or three most important concerns of planners and policy-makers in recent years, and it is definitely the most important transport problem from the perspective of most Americans. That is because the problem is visible and affects so many people every day. The time losses congestion causes are immediate, in contrast to the more delayed impacts of pollution.

The most obvious reason for increased traffic congestion in American cities is the very rapid growth in traffic on a road network that has expanded very slowly in recent decades (Downs, 1992). The supply is simply inadequate to handle the demand, especially under the current policy of underpricing of car use and road access. At such low user cost, car use is bound to be excessive. The increase in traffic congestion has been worst in the suburbs, where the growth in automobile travel has been concentrated, and where development is so sprawled that public transport, walking, cycling and even car-pooling do not offer realistic alternatives to single occupant car use.

Virtually all studies indicate that average speeds on urban roads are falling considerably, especially in the suburbs (Downs, 1992). Moreover, a recent study by the US Department of Transportation (1994) found that the total hours of congestion delay in the 50 largest American cities increased from 8.3 million in 1986 to 10.1 million in 1990, a 22 per cent increase. Remarkably, over half of the nation's total hours of congestion delay were concentrated in five metropolitan areas: Los Angeles, Washington, San Francisco, Chicago and New York. Thus, the problem of traffic congestion is by no means equally serious in all cities. The same study, however, confirms that the total hours of congestion delay in every city has been increasing since 1986, almost regardless of city size and location.

There is some debate whether people are actually spending more time travelling than previously. Gordon *et al.* (1991) argue that people have been shifting their residential and workplace locations in order to reduce their travel distances and thus offset the reduced travel speeds. Examining trend data from the 1977 and 1983 Nationwide Personal Transportation Surveys, the 1980 US Census and the 1985 Annual Housing Survey, they find that average travel times for commuters remained stable or even decreased slightly. The 1990 US Census reported a slight increase in average commuting time from 1980 to 1990 (from 21.7 to 22.4 minutes). By comparison, the 1990 Nationwide Personal Transportation Survey indicates a long-term decrease in the average commuting time, from 22.0 minutes in 1969 to 19.7 minutes in 1990 (Federal Highway Administration, 1993). Although average trip times may not be increasing, the perception among virtually everyone is that congestion has become much worse over the past 10 years (Downs, 1992). Rising expectations may explain some of the increasing dissatisfaction with traffic conditions.

Accidents

Traffic accidents have for many decades been the most important cause of unnatural death in the USA. As shown in Table 10.7, the total number of traffic fatalities has fallen considerably since 1970, and the fatality rates per 100 million km travelled and per capita have fallen even more sharply. That is the result of significant improvements in the engineering and design of cars to be safer for occupants. Nevertheless, with 40 000 traffic deaths and almost 200 000 serious injuries every year, traffic safety clearly

Table 10.7 Traffic fatalities, 1960–92

Year	*Total traffic fatalities*	*Fatalities per 100 million vehicle-km*	*Fatalities per 100 000 population*
1960	38 137	3.31	21.3
1970	54 633	3.08	26.9
1980	51 091	2.09	22.6
1990	44 599	1.30	17.9
1992	39 235	1.09	15.3

Source: US Department of Transportation, *National Transportation Statistics* (1993).

remains a very important problem. Moreover, as cars have become safer for occupants, virtually nothing has been done to protect pedestrians and cyclists, who currently account for almost 40 per cent of traffic fatalities in urban areas (National Safety Council, 1993; US Department of Transportation, 1993).

Energy

From about 1973 to 1983, energy use in transport was at the top of the agenda as a transport problem in the USA. Petroleum shortages, large price increases and the dependency on foreign oil suppliers all led to major efforts to make the car more fuel-efficient. As shown in Table 10.8, those efforts paid off. The average fuel efficiency of new cars almost doubled between 1970 and 1992. At the same time, the real, inflation-adjusted price of petrol has been falling in recent years, and is now at its lowest level in decades. Neither the general population nor politicians currently view energy to be a very important transport problem, although the USA is now more dependent than ever on foreign oil supplies, and a sudden interruption in supplies would provoke a crisis such as in 1973 and 1979. Moreover, many analysts have argued that the USA would never have entered the Persian Gulf War of 1991 except to ensure future supplies of cheap oil. That war obviously had an enormous cost.

Equity

One of the most severe transport problems in the USA is the tremendous inequality in mobility among different socioeconomic groups. The poor,

Table 10.8 Energy efficiency of cars, 1960–92

Year	Average miles per gallon for entire fleet	Average miles per gallon for new cars
1960	14.3	16.1
1970	13.5	15.2
1980	15.5	24.3
1990	21.0	28.0
1992	22.0	27.9

Source: US Department of Transportation, *National Transportation Statistics* (1993).

the elderly and the handicapped make less than half as many trips per capita as the rest of the population (Altshuler *et al.*, 1981). To some extent, their lower journey rates result from less participation in the labour force (and thus fewer commuter trips) or a less active lifestyle. Transport problems, however, also limit their mobility. Due to financial or physical constraints, these groups have less access to automobiles, which have become increasingly essential to meeting travel needs in America's low density, suburbanized metropolitan areas. Indeed, most recreational, educational, shopping and employment sites are almost totally inaccessible without cars. Thus, disadvantaged groups in the USA are yet further discriminated against by a transport system that drastically limits the mobility of anyone without a car.

That severe inequity in mobility has also reinforced the extremely severe problem of racial and economic segregation in American cities. The lack of automobile access has certainly helped keep the poor out of the suburbs. Unlike Europe, most of America's urban poor live in the central cities quite close to central business districts. The spatial concentration of poverty in racially segregated slums has exacerbated the social and economic problems arising from low incomes, creating what some analysts call a 'culture of poverty'. Moreover, the segregation of the poor in inner-city slums has led to discrimination in the provision of public services as well. Educational, recreational, sanitation and health services in slums, for example, are far inferior to those in the affluent suburbs surrounding American cities.

The spatial segregation of different racial and income groups has been crucial in maintaining the highly stratified socioeconomic structure in the

USA (Harvey, 1989; Yeates, 1990). The car-oriented system of urban transport has been the most important factor enabling the spreading-out, distillation, and segregation of different racial, ethnic and economic groups in America's cities. America's social problems get worse and worse as the gap between rich and poor grows and different racial, ethnic and economic groups live and work further and further from each other.

Other problems

The list of problems of automobile use goes on, but many problems are almost impossible to quantify. They include, for example, noise, urban sprawl, the depletion of natural resources and the loss of open space. The World Resources Institute (1992) estimates that the total cost of the social and environmental harm done by automobiles is at least $400 billion per year. The Office of Technology Assessment (1994) of the US Congress estimates the total social, environmental and external economic costs of car use to be at least $447 billion per year, and possibly as high as $899 billion per year if all relevant impacts, both direct and indirect, are included. If those external costs of car use were assessed to car drivers in the form of petrol taxes, it would require a supplemental tax of at least $3 per gallon to internalize those costs. In fact, car drivers in the USA are not required to bear such costs themselves. Thus the huge external costs imply an enormous underpricing of car use in the USA, and an implicit subsidy to driving. The severe underpricing of automobile use in the USA is at the root of most problems of car use.

Financial burden of transport subsidies

The total government subsidy to roads and public transport in the USA grew from about $11 billion in 1970 to $46 billion in 1992. Especially in the current environment of budget cutbacks and fiscal austerity at every government level, that huge public cost of transport finance has become even more of a burden. As discussed in the next section, the most recent federal legislation permits large increases in subsidies for ground transport. It was justified mainly on the basis of job creation during the 1990–92 economic recession. The Republican takeover of both houses of Congress in 1995 – together with the greater fiscal conservatism of the Clinton Administration – virtually guarantees that any actual subsidy growth in coming years will be far smaller than the amounts authorized by law until 1997. Indeed, subsidy cutbacks now seem quite likely.

From an economist's point of view, the problem is not simply the size of the subsidies but also the wasteful impacts subsidies seem to have. Many urban transport projects, for example, cannot possibly be justified on the basis of transport needs or total benefit, but result from political considerations such as logrolling and pork barrel politics (Altshuler *et al.*, 1979). Moreover, the very availability of large subsidies appears to have encouraged lower productivity, higher per unit costs, overstaffing and the overdesign of projects (Pucher *et al.*, 1983). Finally, federal subsidies in particular come with many wasteful, time-consuming regulations that have inflated costs, biased state and local decision-making, and greatly politicized the entire transport planning process.

URBAN TRANSPORT POLICY

The federal government has had an overwhelming influence on urban transport policy in the USA. Not only has it funded a large proportion of public costs, its various matching programme have so strongly biased the decisions of state and local governments that federal policy has ultimately shaped policy at all government levels.

Three policy eras

Overall, policy towards urban transport can be divided into roughly three periods between the Second World War and the present (Altshuler *et al.*, 1979). Prior to 1970, the main emphasis was on road building as the main approach to deal with what was then viewed as the main problem: increased congestion due to the rapid growth in car ownership and use. Beginning in 1921, the federal government helped finance highway construction. During the 1950s and 1960s, however, the federal government greatly increased its subsidies to road construction, offering a variety of different subsidy programmes, with the highest matching rates for limited-access superhighways, and successively lower matching rates for primary, secondary and local roads (Weiner, 1992). State and local governments, of course, provided complementary funds and were responsible for suggesting the projects undertaken, supervising the construction, and maintaining the roads after they were built. The total subsidy to roads from all government levels grew steadily and reached \$20.8 billion in 1970, of which \$10.2 billion was financed by road user charges, leaving a net subsidy of \$10.6 billion financed out of general government revenues (see Table 10.6).

At the same time that massive subsidies were channelled into road construction, urban public transport was almost completely neglected. Public transport services became ever less profitable and then ever more unprofitable during the 1950s and 1960s. Some private firms abandoned service altogether and others were taken over by city governments. Overall, however, state and local government aid was minimal, and federal aid did not begin at all until the mid-1960s (Weiner, 1992). Gradually, of course, public transport subsidies did increase, but by 1970 the total subsidy from all government levels was only $518 million (less than one-twentieth the net subsidy to roads). As a consequence of financial difficulties, public transport systems were again and again forced to raise fares, curtail services, neglect maintenance and forgo any sort of expansion, rehabilitation or modernization. Ridership plummeted until the mid-1970s, falling to only about a third of its 1945 level.

The decade of the 1970s witnessed a profound shift in federal transport policy. There was growing public awareness of the many negative impacts of automobile use and highway construction. One problem after another was added to the list of car-related ills: central city decline, traffic accidents and injuries, mobility problems of the poor, congestion, air pollution, destruction of inner-city neighbourhoods, energy waste and lack of accessibility for the elderly and handicapped. Largely as a reaction to these negative impacts, the rate of road construction slowed down considerably, and government funding for public transport skyrocketed. Indeed, the total subsidy from all government levels rose 14-fold, from $518 million to $7139 million. By comparison, the total subsidy to roads only doubled (see Table 10.6).

Unfortunately, much of the increased funding for public transport was wasted. Political considerations outweighed economic rationality. Too much money was focused on fancy new heavy rail systems and too little money on rehabilitation and modernization of old systems. Moreover, most service expansion (including new bus routes) was in low density suburban areas, while central city services were often neglected. The new public transport services were much more expensive than older ones and, with few exceptions, they generated disappointingly little new usage. Finally, the sudden large increase in subsidies led to rapid escalation in per unit costs. Both labour costs and capital costs grew much faster than inflation (Pucher *et al.*, 1983).

Not only did subsidies increase from 1970 to 1980, but the federal government doubled its share of subsidy funding (from 26 per cent to 54 per cent), and almost every urban area created special region-wide public transport districts and dedicated special taxes exclusively to finance the necessary subsidies. Finally, the decade of the 1970s witnessed a complete public takeover of the industry so that, by 1980, virtually all public transport service was publicly owned and operated.

Partly due to the rather disappointing results of increased subsidies and partly out of general Republican opposition to federal subsidies, the Reagan and Bush Administrations repeatedly tried to extract the federal government from the enormous public transport subsidy programme that had been built up during the 1970s. Those efforts were not successful because the Democratic-controlled Congress vigorously resisted subsidy cutbacks. Nevertheless, Reagan and Bush managed to reduce sharply the proportion of the total subsidy funded by the federal government (from 54 per cent to 24 per cent) and to curtail somewhat its subsidy contribution. State and local governments took up the slack left by federal subsidy reductions and provided substantial increases in their own contributions. The net result was continued growth in total subsidy until 1990, but the average annual rate of growth slowed down dramatically.

Federal transport legislation

Underlying all the trends in government policies have been a series of federal transportation laws, which have revised funding and regulations for both public transport and roads. Moreover, numerous federal laws pertaining to safety, the environment, energy, civil rights and the handicapped have had important implications for transport as well. A detailed description of the most relevant federal laws can be found in Weiner (1992).

In general, the federal transport laws have provided increased funding for both public transport and for highways. At the same time, federal regulations have burgeoned, and now cover virtually every aspect of transport planning, design, and finance. Most federal regulations have been well intentioned, aimed at achieving various social and environmental goals. Unfortunately, they have also greatly exacerbated cost inflation of transportation projects. Thus, increased federal aid has been a mixed blessing.

The most recent federal transport legislation is the culmination of trends that had been developing for at least two decades, featuring multimodalism and flexibility in funding use between capital and operating expenditures as well as between roadways and public transport. The Intermodal Surface Transportation and Efficiency Act of 1991 (ISTEA) authorizes \$155.3 billion in federal funding of highway and public transport projects from 1992 to 1997 (Weiner, 1992). Of that total, \$31.5 billion is explicitly for public transport. That amounts to an average of over \$5 billion a year, almost 50 per cent more than the \$3.3 billion per year in the late 1980s. Moreover, the law permits the option of using an additional $58 billion in federal transportation funds for public transport, bicycle and pedestrian facilities (US Congress, 1991; Weiner, 1992). Of that $58 billion, $29.9 billion is explicitly designated as flexibile funding. The Surface Transportation Program of ISTEA authorizes $23.9 billion in block grants to states and urban areas that can be spent on highways, bridges, public transport, car-pool projects, bicycle and pedestrian facilities, safety, traffic management, pollution control and planning. State and local governments and regional metropolitan planning agencies determine how the funds are spent. Likewise, the Congestion and Air Quality Improvement Program of ISTEA, which is authorized at a level of $6 billion over six years, can be used at the option of state and local governments for a broad range of transportation programmes, provided that they can be shown to improve air quality and reduce congestion. Finally, states have the option of shifting an additional $28 billion – 50 per cent of their federal highway funds and 40 per cent of their federal bridge funds – to projects that would improve public transport, car-pooling, cycling or pedestrian facilities.

Not only was more funding supposed to be made available for public transport, but the federal matching rate was equalized at 80 per cent for most federally assisted capital investments both for public transport and for highways. Interstate highways still get funded at the 90 per cent federal match level, but that is somewhat offset by an equally generous 90 per cent match for public transport vehicle costs incurred to comply with federal environmental standards (set by the 1990 Clean Air Act) and to provide mobility to the handicapped (as mandated by the Americans with Disabilities Act).

ISTEA is still in the first stages of being implemented, but it is already clear that actual federal funding is falling far short of the subsidy levels authorized by Congress. In the 1992–93 fiscal year, the actual subsidy appropriation for public transport was $3.8 billion, only 71 per cent of the ISTEA authorized subsidy level. In fiscal year 1993–94, actual appropriations for public transport were somewhat higher (about $4.6 billion) but still only 86 per cent of ISTEA authorized levels (American Public Transit Association, 1994). In the coming years, public transport subsidies almost certainly will fall. As of mid-1995 both the Clinton Administration and the Republican Congress are proposing substantial cuts in subsidies. Indeed, the Republicans want to eliminate operating subsidies altogether. The outcome of the budget deliberations is uncertain. It is clear, however, that the subsidy programme is falling victim to the federal government's budget crisis, the Republican takeover in Congress and the Clinton

Administration's corresponding shift to the right and much greater fiscal conservatism.

In addition to diminished federal funding for public transport, state and local governments have been deciding to allocate almost all of their federal transport block grants and flexible funds for highways and not for public transport. Overall, the actual subsidy levels for public transport represent only about a fifth of the total ISTEA funding potentially available for public transport (including all the various flexible and transferable funds). In sharp contrast to the hopeful prospects for increased public transport funding when ISTEA was enacted in 1991, it now seems certain that the coming years will bring substantial subsidy reductions, thus exacerbating the financial crisis of public transport in the USA.

Road construction and travel demand management

As travel demand has burgeoned in recent decades, road supply has stagnated. Between 1975 and 1992, urban road mileage in the USA expanded by only 23 per cent, compared to an 87 per cent increase in vehicle miles of travel in urban areas (Federal Highway Administration, 1985 and 1993). Community opposition, environmental objections, development moratoria and funding shortages all contributed to the slow-down in new highway construction. In addition, many transport analysts had criticized the massive road construction programmes from 1960 to 1975. Not only was new highway construction found to be socially and environmentally harmful, but its short-term improvements in capacity were quickly offset by the long-term increases in travel demand caused by the extensive suburbanization new roadways stimulated. Although the 1991 federal transportation law (ISTEA) authorizes increases in federal funding for road construction, environmental and community opposition to almost all proposed expansion projects remains intense. It will probably be possible to make a few selective additions to the urban road network, but it is virtually inconceivable that capacity expansion will be the main solution to the congestion problem.

Increasingly, academic policy analysts as well as government policymakers are turning to demand management to deal with congestion. Some demand management techniques are already in use in American cities. For example, over 20 cities have established express lanes for HOVs on about 60 key arterial freeways (American Public Transit Association, 1993). A few cities have also experimented with selective ramp metering on freeways to give priority to HOVs. Moreover, some cities have transformed one or two of their central streets into pedestrian or transit malls, with only limited access for cars and trucks. Such malls, however, are far less extensive and less coordinated than the large car-free zones in most European cities. Finally, a few cities are even experimenting with limiting the supply of parking in central city areas instead of requiring large amounts of parking as in the past.

Measures aimed at reducing car traffic – instead of accommodating it – have become the preferred solution among transport planners, and are increasingly feasible politically because they have beneficial environmental impacts, require only minimal funding and can be implemented quickly. Moreover, the Clean Air Act Amendments of 1990 virtually require the adoption of such transportation demand management strategies in those metropolitan areas that exceed the maximum air pollution levels allowed by the US Environmental Protection Agency (Weiner, 1992). Urban areas in non-compliance with Environmental Protection Agency standards must establish specific plans for reducing vehicle miles of car travel and discouraging single-occupant use. State and local planners are required to assess the pollution-reduction potential of improved public transport, HOV lanes, fringe parking facilities, car-pooling and van-pooling programmes, transport pricing measures and improved bicycle and pedestrian facilities. In urban areas with severe pollution (including many large American cities), all firms with more than 100 employees must establish programmes to increase employee vehicle occupancy for work-related trips by at least 25 per cent above the average for their urban area. State and local government officials are responsible for monitoring and enforcing the plans submitted by individual firms and must impose fines on those firms failing to comply. The US Department of Transportation is empowered to withhold federal highways funds from metropolitan areas that do not enforce the new regulations (Weiner, 1992).

Although the Clean Air Act Amendments of 1990 have been widely touted for the far-reaching impacts they could have on travel behaviour, their actual impact is likely to be minimal. Southern California has had a similar law (Regulation 15) in effect for the past four years, and a recent study of its impacts found only a small overall reduction in single-occupant car use (0.4 per cent decline in car kilometres driven), and no increase at all in public transport usage (Wachs, 1993). Moreover, it is highly questionable whether the federal government will really withhold highway funds from those states failing to comply. As in the past, political pressures will probably succeed in forcing exceptions, delays and lowering of standards.

Congestion pricing continues to be a popular notion among some transport analysts, especially economists, but in the three decades since it was first proposed it has made virtually no headway. The 1991 federal transportation act may revive interest as it provides funding for five experimental congestion pricing projects (Weiner, 1992). Perhaps the experiments will be successful, but it seems inconceivable that Americans would accept congestion pricing for widespread implementation. As a practical matter, it faces insuperable political opposition in the USA (Giuliano, 1992). Even small increases in the existing petrol tax are vigorously attacked by industry lobbyists and consumer groups. Congestion pricing, by comparison, would be viewed as a totally new kind of taxation, an infringement on personal freedom and an inequitable rationing of mobility. The congestion pricing experiment may yield some interesting data for economic analysis, but it probably will not provide a model that can be widely followed.

Improving transport equity

Most of the transport policies in the USA dealing with equity have focused on the special mobility problems of the elderly and handicapped (Pucher, 1982; Weiner, 1992). The federal government has mandated a wide variety of programmes to expand transport services for them. For example, new buses must be equipped with wheelchair lifts, and rail stations must have ramps and lifts for the handicapped. Moreover, public transport systems must provide substitute door-to-door dial-a-ride services (in minibuses or vans) in those corridors where their regular fixed-route services are not fully accessible to the handicapped. Finally, a variety of federal social service programmes provide funding for completely separate minibus and van services for the elderly and handicapped. Unfortunately, it has not yet been possible to integrate these separate services to provide a coordinated system of special transport services (Rosenbloom, 1992).

The considerable political power of organizations representing the elderly and handicapped explains the extensive federal involvement in transport policy for those groups. By contrast, almost nothing has been done to solve the enormous mobility problems of the poor. There have been a few demonstration projects providing bus or van services between central city slums and suburban job sites. Most government efforts, however, have been limited to preventing deliberate racial discrimination in service and fare policies of public transport systems.

One of the general rationales for subsidies to public transport in the USA has been the assumption that most public transport riders are poor and thus are the main beneficiaries of government subsidies. In fact, several studies have shown that most public transport riders are not poor and that most of the poor never use public transport. To make matters worse, the distribution of subsidies is inequitable, with the largest subsidies going to rail services used by affluent riders, and the smallest subsidies going to inner-city bus services used most by the poor (Pucher, 1982). Nevertheless, poor residents of America's inner cities are far more dependent on public transport for their mobility needs than affluent suburbanites. Thus, to the extent that general subsidies to public transport are necessary to finance comprehensive metropolitan region-wide services, such general government support for public transport might be viewed as indirectly – and perhaps inadvertently – mitigating the mobility problems of the poor.

Regulation of car manufacturers

The federal government's main approach to dealing with the social and environmental costs of automobile use has been to require manufacturers to produce successively safer, cleaner, quieter and more energy-efficient cars. There are a myriad of very specific standards that new cars must meet. Each manufacturer, for example, must achieve a minimum average fuel efficiency, weighted by fleet composition. Emission standards for HC, CO, NOx, particulates and lead must be met for all new cars. Those standards have become stricter with each successive Clean Air Act. Similarly, safety standards include requirements for seat belts, air bags, shatterproof glass, crash-resistant bumpers and fuel tanks, fire retardant upholstery and many other safety features.

A more recent technogical approach to mitigating the problems of automobile use is the IVHS (Weiner, 1992). In its first stages, the programme aims to provide car and truck drivers with specific, up-to-date information on traffic conditions, optimal routing and potential safety problems. In more advanced stages, the aim is to use computer information and guidance technology to route and operate vehicles in a way that would minimize travel time and increase safety. As touted by its proponents, IVHS technology would eventually take over much of the task of driving, and thus presumably avoid the problem of human error. Although IVHS has much appeal among transport engineers, computer specialists and consulting firms, it is debatable whether or not IVHS can really solve any transport problems. Environmentalists, urban planners and proponents of non-automobile modes have attacked IVHS as no more than a source of lucrative contracts for consultants and a waste of scarce public funds. Nevertheless, the 1991 federal transportation law (ISTEA) authorizes

$659 million over six years for IVHS research, 79 per cent of all federal funding for transport research (US Congress, 1991).

Whatever their limitations, technological approaches to dealing with the problems of automobile use have been especially appealing to politicians and, of course, to the public as well. Improving the design of cars appears to have only benefits for the user, with the costs of design changes hidden in the purchase price. Moreover, technological improvements avoid the necessity to change travel behaviour, location patterns and lifestyle. Finally, most cost-benefit studies suggest that, at least for most changes in car design, the overall benefits have, in fact, exceeded costs.

Unfortunately, not all harmful impacts of car use can be so conveniently mitigated through painless technological change. As already noted, for example, almost 40 per cent of traffic fatalities in cities are pedestrians and bicyclists who hardly benefit from measures to increase the safety of automobile occupants. Similarly, problems of equity, neighbourhood deterioration, urban sprawl and congestion are not likely to be solved through technological advances in car design.

Housing and land use policies

The American public sector has made few efforts to control urban sprawl or to encourage more compact development (Altshuler *et al.*, 1979). For the most part, development in the USA arises haphazardly as private developers, builders and land speculators try to maximize their profits with little regard, if any, for social and environmental consequences. Implicitly, however, the public sector has strongly encouraged low density suburbanization.

The federal government has offered mortgage insurance and various tax advantages for home owners since the Second World War. Perhaps most significant, mortgage interest and local property taxes on single-family, owner-occupied houses can be deducted from federal income taxes. In 1990 alone, the total of such home owner deductions amounted to $232 billion (US Department of Commerce, 1993). Those deductions, together with various home owner exemptions from capital gains tax on home sales, result in forgone federal tax revenues of $75 billion per year. Over the course of the past five decades, the cumulative implicit subsidy to home owners has amounted to hundreds of billions of dollars. That huge subsidy has unquestionably biased residential choices towards low density suburbs and, by subsidizing residential suburbanization in this manner, federal policy in the USA has promoted the very kind of low density sprawl that is almost impossible for public transport to serve effectively.

State and local governments have provided the essential capital infrastructure and public services that new suburban development requires: sewers, water lines, fire and police protection, and schools. The necessary public expenditures have far exceeded the additional tax revenues that suburban development has generated. Moreover, some studies find that the fragmented, uncoordinated structure of local government in American metropolitan areas, by its very nature, has subsidized suburban residents and firms at the expense of their central city counterparts (Advisory Commission on Intergovernmental Relations, 1984; Neenan, 1972; Yeates, 1990; Heilbrun, 1987). Finally, the public construction of suburban roads and limited-access highways within metropolitan areas has obviously spurred the decentralization of population and employment (Schaeffer and Sclar, 1980).

Canadian and European land use and housing policies have either explicitly encouraged compact development or prohibited certain low density forms of urban sprawl. Moreover, they have generally restricted the amount of land near cities that can be developed at all for residential or commercial uses. That, of course, limits the supply of developable land, raises its price and encourages higher density development. Not only have land use regulations in the USA been lax, thus permitting sprawled, low density development; policies at every government level have provided strong financial incentives to suburbanize. Thus, it is not surprising that American suburbanization is the most extensive and lowest density of any country in North America or Europe.

As mentioned at the outset, the extremely low density suburbs that dominate American metropolitan development are almost impossible to serve effectively with public transport, thus ensuring the dominance of automobile travel for many decades to come. Admittedly, both automobile use and suburban single-family homes have great inherent appeal to American consumers. There can be no question, however, that public policies in the USA not only accommodates those consumer preferences but actually makes automobile use and suburban living possible at artificially low prices. Subsidies, tax deductions, road construction, suburban infrastructure provision, and the failure to internalize the social and environmental costs of auto use: those are all examples of the crucial role the American public sector played in establishing the automobile as the overwhelmingly dominant mode of urban transport, and the single-family suburban house as the dominant form of residence.

CONCLUSIONS

The car's dominance of passenger transport in American cities is certain to continue over the coming decades. Even huge increases in public transport subsidies were incapable of reducing the car's share of total urban travel from 1970 to 1990. On the contrary, car modal split rose further, and public transport modal split continued its decline of over four decades. Public transport subsidies will probably decrease in the coming years. Thus, service expansion and quality improvement will probably be less than during the past two decades, and fare increases are likely to be larger. It will not be easy to maintain current levels of use, let alone increase them.

The continuing rapid decentralization of population and employment in American metropolitan areas is the main obstacle to improving the performance of public transport and the main guarantee for ever greater automobile dependence in the future. Even the best of public transport service cannot really compete with the automobile in America's very low density, polycentric suburbs.

Another insuperable obstacle for public transport is the serious underpricing of automobile ownership and use in the USA. The subsidy to car use is mostly indirect but it is enormous, amounting to over $400 billion per year according to the World Resources Institute (1992) and $447–899 billion per year according to the Office of Technology Assessment (1994). The provision of free parking, the failure to charge car users for social and environmental costs, and direct subsidies to road users have made automobile ownership and use so inexpensive that they are virtually irresistible. That puts public transport at an impossible competitive disadvantage.

Now that the vast majority of Americans own cars and live in the suburbs, there is overwhelming political support for continuing the public policies that have favoured automobile use and suburban living in the past. As those public policies persist, it is highly likely that American cities will become even more car-dependent than they are now.

The best that can be done is to minimize the damage done by such excessive automobile use by reinforcing government efforts to require technological improvements to cars that increase safety and fuel efficiency while reducing noise and air pollution. Such improvements will probably prevent a worsening of America's urban transport crisis, but they do not get at the heart of the problem, which is the underpricing and oversubsidization of both car use and low density suburban living.

11 Dealing with the Urban Transport Crisis: Comparative Policy Evaluation

In the preceding 10 chapters, we have documented the trend towards increased car ownership and use throughout Europe and North America. Although the USA led the way in the shift from public transport to the car, Canada and Western Europe have been following the American example for the past two decades, and Eastern Europe is now following the example of Western Europe. At the same time, the use of public transport, walking and cycling have been declining as modes of urban travel. There remain significant differences in travel behaviour between Europeans and Americans, and also between Americans and Canadians, with the USA still by far the most automobile dominated country in the world. The differences, however, become smaller with each passing year.

Just as car use has spread from the USA to the rest of the world, suburbanization is also becoming a world-wide phenomenon. Although the extraordinarily low density, unplanned suburban sprawl around American cities probably represents the extreme form of urban decentralization, European cities are rapidly decentralizing as well, albeit at much higher densities than American suburbs.

Urban decentralization has greatly increased travel distances and has reduced the relative importance of trips to and from the city centre, which public transport serves best. Travel between and within suburbs, on the other hand, is growing fast in all American, Canadian and European cities, and it is precisely this sort of trip pattern for which the car is a virtual necessity. Such suburban trips are usually too long for walking or cycling, and they do not generate high enough travel volumes in route corridors to make public transport economically feasible. Thus, suburbanization has sharply reinforced the trend towards ever greater use of, and dependency on, cars.

Rapidly growing car use has been the primary source of urban transport problems in Europe and North America. Of course, public transport also causes some air pollution, noise and accidents, but the automobile is main culprit. Not only does the automobile produce more road congestion, pollution, noise, traffic accidents and energy use per passenger kilometre of travel but, together with the suburban development it promotes, the automobile generates much more travel per capita than public transport, and thus causes a great deal more social and environmental harm.

As described in detail in the preceding chapters, the nature and extent of the urban transport crisis varies from one country to another. Congestion, air pollution, safety and finance are the most widespread problems. Energy use, suburban sprawl and inadequate accessibility for certain social groups are severe problems in some countries but barely noticed in others. Moreover, the objective extent and subjective evaluation of transport problems can vary dramatically over time. Whatever the variations from one country to another, urban transport problems plague cities in all countries. They have been getting worse in recent years, and most projections indicate that this trend is likely to continue in the years ahead.

INADEQUATE POLICY RESPONSES

Unfortunately, public policies towards urban transport in most countries have not been very effective at solving urban transport problems and, in many instances, they have greatly exacerbated the situation, leading to a transport crisis. In general, policies have tended to follow social and economic trends as well as public opinion. Consequently, they have been car oriented, as we have shown in this book. For instance, in the 1950s and 1960s, there was an important political and social consensus around the automobile. At the time, the car was considered a panacea for transport problems; it was a symbol of wealth and social progress and as such was not constrained. Indeed, public authorities adopted policies that favoured the car by building more roads and more parking facilities, and by making automobile ownership and use easier for all. This changed slightly in the 1970s with the awareness of the enormous economic, environmental and social cost of automobile dominance. Since then, public policies have been partially reoriented but never completely reversed even in the less car-oriented countries. Moreover, these policies did not have a significant impact on the general trends described in this book. For instance, although large public transport subsidies have succeeded in keeping public transport alive in most urban areas, in many cases they have not been able to increase public transport patronage to any significant degree, and in most countries they have not been able to prevent a decrease in the modal split share of public transport.

Pro-automobile policies have obviously been implemented to the detriment of other transport modes although the extent of the favouritism has varied considerably from country to country, with nations like the Netherlands attempting (with some success) to maintain public transport as well as cycling and walking. In most countries, however, these transport modes have been neglected, leading to their sharp decline in the modal split; in some urban areas, they have almost disappeared. Walking and cycling are surely the most economical, most energy efficient and most environmentally friendly modes of urban transport: yet they have suffered greatly from the massive decentralization of urban areas, which puts more and more destinations outside their reach. Moreover, most countries have neglected to provide even the minimal levels of public investment required for the safety and convenience of pedestrians and cyclists.

Although public transport has not been neglected to the same extent, the lack of adequate funding in that field in the last three decades helps explain the degradation of public transport quality in most European and North American urban areas. Huge subsidies have been injected into public transport in most countries, including the USA, but those funds have not succeeded in producing high quality public transport networks. Thus, in most large urban areas, public transport infrastructure remains inadequate. In Paris, for example, the RER line A is overcrowed and the quality of service is deteriorating on a number of metro and rail lines (more delays, crowded carriages), to such an extent that it is considered the major cause of the recent significant loss of ridership. The same situation applies in most European cities. In London, for instance, a large part of the BR network is already saturated, and in the Randstad, the NS network operates over capacity. Insufficient funding is also evident in the delays in extending metro lines in many urban areas (Amsterdam, Hamburg, London, Milan, Munich, Rome, Rotterdam). So far, accessibility by public transport has not improved over the years in spite of huge investments and subsidies.

This is why public transport policies can be said to have failed to create an attractive alternative to the automobile. In some countries (France, Italy the United States) public transport is still considered a mode for the 'others': the poor, those who do not or cannot own a car. If public transport is losing ridership in many countries and cities, it is because public transport networks and services are less and less able to respond to the mobility patterns of the population so that more and more people are having to rely on the automobile to travel. This is especially true in the suburbs where public transport networks are rare and infrequent services the rule. This inability of public transport policies to create a convenient alternative to the private car is all the more dramatic when the proportion of households depending on public transport remains high (between one-tenth and one-third of total households) in spite in the dramatic increase in car ownership.

By inducing excessive car use, pro-automobile policies have necessitated large investments and operating subsidies for public transport. Subsidies for public transport have become so large that, in some countries, deregulation and privatization policies have been launched to reduce subsidies and increase productivity. The experience of Great Britain confirms that deregulation and privatization can reduce costs and subsidies, but only at the expense of higher fares, reduced services and declining public transport ridership. Moreover, full deregulation makes coordination and long-term planning of public transport services more difficult, since it permits numerous operators with strong incentives to work against each other instead of with each other. The ease of market entry and exit under full deregulation also makes route structure and service frequency less stable. At any rate, full deregulation and privatization are certainly not the panacea for public transport's problems.

More limited types of deregulation and privatization, such as the franchising system through competitive tendering implemented in Scandinavian countries in the 1980s and 1990s, seem more promising. Recent studies (Andersen, 1993) have shown that such measures have been successful in reducing costs and subsidies significantly without creating the drawbacks seen in Britain. Control over fares and services has remained in the hands of authorities who ensure the coordination of lines and networks.

Technology has been a constant focus of public policies almost everywhere in the firm belief that technological advances in car design, electronics, telematics, information systems and automation would help solve the most common transport problems. Although it is true that technological advances in design have mitigated somewhat the pollution, safety and energy problems of automobile use, these benefits have been largely offset by the rapid increase in the total amount of travel. There are, in fact, limits to the technological responses to the urban transport problem. Some problems, such as suburban sprawl and social inequities in accessibility, are simply not amenable to technological solutions. Other problems, such as congestion, might be alleviated through the proposed smart highway/smart car programmes, but such computer optimization is very expensive and is more likely to provide temporary congestion reduction rather than long-term relief.

The significance of technology programmes and innovations to respond to some transport issues may be better understood by studying their economic and political role. For car manufacturers as well as public transport manufacturers and operators, technological innovation is obviously regarded as a factor which will increase the product's attractiveness to potential buyers. For politicians, it may be a way to avoid critical decisions. For instance, it is politically easier to fund telematic systems which would inform passengers of the exact arrival times of buses or trams (most likely delays) than to implement segregated rights of ways in the streets. Reserved lanes for public transport could ensure a more reliable service at a lower cost; but such a decision is likely to entail conflicts with automobile drivers and is therefore not often implemented, although frequently discussed.

Finally, transport policies have always been sectoral. In most cases, transport has been considered a separate element in the urban system. Transport policies have not been designed as part of urban planning strategies, notably those regarding the location of residences and activities. This can be illustrated by the inadequate public transport service to the two major Paris airports (Orly and Roissy). Orly airport was built in the 1960s, but a connecting rail line (Orlyval) was only constructed in 1991. Roissy airport was built in the 1970s, but the direct connection with the city centre was only established in December 1994.

The absence of a link between transport and urban planning is to be found within the transport sector as well. Indeed, highway policies and public transport policies have evolved along separate, often conflicting, lines. This situation is the result of political and financial mechanisms which have created a strong autonomy for each mode. For example, in most countries, public transport and highway policies rest with many different authorities (municipalities, joint authorities, states, provinces). In most cases, these authorities do not coordinate their actions and policies with each other. Second, most countries have provided specific funding for each mode in recent decades. For instance, the French payroll tax may only be used for public transport projects. In the United States, the Highway Trust Fund (created in 1956) has long been used for highway projects. In the 1960s and early 1970s, huge sums were available to build highways when public transport networks were collapsing. The diversion of a portion of highway monies to public transport, although partially authorized in the mid-1970s, was seldom carried out. This lack of funding flexibility was a key problem in federal transport funding until 1991, when the new transport act ISTEA was passed. Even since then, actual flexibility has been much less than allowed by law, since states have been continuing to use federal funds along historical, pro-highway patterns. In the UK, the same situation exists; the possibility of using highway funding for public transport has only emerged very recently, in fact since 1993 with the 'package approach' proposed by the DoT (see Chapter 7) which is considered a significant change in the mode of funding transport infrastructure. In short, in all cases, these financing mechanisms have not allowed for a true choice between highway and public transport schemes. Highway projects were generally favoured because funding for them was greater and easier to use. Moreover, the opportunity to choose between a highway project and a public transport project was seldom available.

IMPROVING TRANSPORT POLICY: WHAT CAN COUNTRIES LEARN FROM EACH OTHER?

If we establish a reduction in the dominance of the car as a general objective of transport policies, then the achievement of such a goal requires an integrated and balanced transport system. First, urban transport policies must be directly coordinated with land use policies. Second, they must focus on the journey – and not the mode – as the basis of analysis and measures. Although these two criteria for transport policy may seem evident, they are seldom fulfilled.

A few countries, such as the Netherlands, Germany and Switzerland, have in fact been rather successful in relating land use policies and urban transport policies. They have implemented extremely strict land-use controls, thus sharply limiting sprawl and encouraging much more compact suburban development than in the USA or the UK.

In the transport domain itself, policies must not be sectoral, segmented by mode. Instead, they should concentrate on entire journeys from origins to ultimate destination and view these as chains of trips and combinations of modes. By so doing, the focus of transport policies would be the management of journeys and would emphasize 'forgotten or ill treated elements'. More attention must be given to transfer points (park-and-ride facilities, transfer among different public transport routes and modes and also between the bicycle, walking and public transport), 'terminal' trips (for instance the connection between the origin or destination of each trip and the principal mode used), and fare or charge integration between among several modes. The Netherlands, Germany, Belgium, and Switzerland have been at the forefront of such coordination efforts. Bicycle parking, for example, is available at virtually all railroad stations, with discount fees for bicycle lockers used by public transport passengers. Moreover, all four countries have carefully integrated different modes and routes of public transport in order to facilitate transfers for passengers. Fare structures have also been fully integrated, thus making the use of different public transport modes and routes much easier than formerly. In short, the full coordination of public transport services and fare structures is not simply a theoretical ideal; it has actually been achieved in these countries of Northern Europe. Unfortunately, such advances in public transport planning have not yet crossed the borders into Southern Europe.

Implementing truly intermodal policies requires a political and technical organization to coordinate all the individual elements of the transport system. One of the most important improvements, therefore, must be in the institutional domain. It would focus on the establishment of an area wide authority in charge of the entire urban transport system (all modes, parking and highways included) with control over land use policies as well. This sort of authority does not exist anywhere. However, some countries have been able to build area-wide institutions responsible for public transport such as the public transport federations (*Verkehrsverbund*) in Germany, Switzerland and Austria. Institutional coordination is essential in the adoption and implementation of integrated transport policies, and the long German experience in this respect should be closely studied by countries like Italy which have not been able to create such authorities in most of their large urban areas.

In virtually all countries, the transport system is unbalanced because it largely depends on one mode: the automobile. Improving transport policies therefore means shifting the balance from the car to other modes. This can be done by putting more constraints on automobile use and by making public transport, cycling and walking more appealing to the urban population.

One important policy mistake has been the provision of ever more motorized transport capacity at low, subsidized user prices, which has encouraged more travel, accelerated urban decentralization, generated yet more travel and stimulated yet further investment in transport capacity. With each stage in this vicious circle, the social and environmental problems of urban transport have worsened. Congestion has increased, not decreased. Similarly, environmental degradation, loss of open space, accessibility problems for the disadvantaged and energy use have become increasingly troublesome aspects of the urban transport system. Consequently, the pricing of automobile use at its full economic, social, and environmental costs must be regarded as a long-term objective if any improvement in urban transport is to be achieved. It would help reduce use of the car, especially in those circumstances where it is most harmful: in large city centres, in congested corridors at peak hours, and in particularly polluted urban areas. Moreover, full-cost automobile pricing would encourage a shift to alternative modes, such as public transport, walking and cycling.

However, such a policy would have to be gradually phased in over the long term because too sudden a policy shift might have harmful, short-term social effects. For instance, any increase in the cost of automobile ownership and use would be more a financial burden for the poor than for the affluent (relative to their incomes).

Increasing the cost of automobile use, however, may not be sufficient, and the social and financial capacity of automobile owners to pay more in order to be able to drive must be borne in mind. In this respect, any policy which limits itself to financial measures is bound to fail. Automobile use must also be restricted by non-financial auto-restraint policies, such as the ones introduced in the Netherlands, Germany or Switzerland (such as traffic-calming areas, pedestrian zones, bus lanes, parking restrictions, and bicycle lanes).

A crucial goal is to offer a true alternative to the automobile: that is, a public transport system which would allow a far higher level of accessibility than is currently possible without a car. This requires more integrated policies as well as the political will to invest more in public transport, cycling and walking facilities, and the social acceptance of such a move.

A few countries have been exemplary in their efforts to revive walking and cycling as viable alternatives to motorized transport. The Netherlands has been at the forefront of such efforts, and has been closely followed by many cities in Germany, Switzerland, Austria and Scandinavia. The establishment of extensive pedestrian zones, wider walkways, integrated networks of bikeways and bike lanes, traffic calming, and the enforcement of traffic rights for pedestrians and cyclists could improve the quality of life in urban cores and residential neighbourhoods everywhere. Various studies have documented the resulting reductions in noise, air pollution and traffic accidents. Further evidence of the success of such policies is that the proportion of urban travel undertaken by walking and cycling has risen to over one-quarter of all trips in most cities implementing them, and in some cases has even reached half of total travel. Of course, walking and cycling cannot possibly serve all travel needs but, where feasible, non-motorized modes should at least be given a chance through a supportive public policy.

Much can be done to enhance the attractiveness of public transport as an alternative to the car. Germany, the Netherlands and Switzerland in particular, have led the way in coordinating fare structures, timetables, routes and different modes of public transport. By integrating services more effectively, they have improved the competitive position of public transport *vis-à-vis* the automobile.

The attractiveness of public transport could also be improved by giving priority status to public transport in urban areas: for instance, by the widespread implementation of reserved bus lanes and priority traffic signals for buses and trams. Experiments with such measures in numerous German, Swiss and Dutch cities have shown that decongesting the rights of way of public transport vehicles can greatly increase their average travel speed, thus significantly raising their attractiveness to users. Moreover, since lanes would be removed from car use and cars would also have to wait longer for green lights at shared intersections, the average speed of the car would decline, thus further discouraging car use and encouraging public transport use instead.

Such transportation system management obviously would improve the performance of public transport. Nevertheless, such low-cost measures must be complemented by large investments in public transport infrastructure in the most heavily-travelled corridors of large cities. Moreover, new forms of public transport service must be developed and coordinated to serve the increasingly important suburb-to-suburb trip, where trip volumes are too low to justify large investments in rail systems. Of course, rail improvements such as metros and LRT may indeed be appropriate on suburb-to-city centre routes, but not for most suburb-to-suburb routes. Lower-density areas require a lower-cost, low-capital, flexible approach.

Van services, minibuses, shared ride taxis and various other forms of paratransit are surely the most appropriate public transport systems for the suburbs. These more flexible, more demand-responsive systems were developed in the 1970s and 1980s in the USA, France and the Netherlands, but the many experiments of those earlier years have not been widely adopted anywhere. Automobile ownership and use has remained so inexpensive that virtually no form of public transport can compete with the car for suburban travel needs.

The range of transport policies suggested above would surely produce a more balanced and integrated transport system, which would offer a really feasible alternative to car use. Unfortunately, their introduction seems unlikely in most countries since they call for a true revolution of travel behaviour in our societies which cannot be expected in the coming years.

CONCLUSION

Powerful economic lobbies and political pressures hinder the adoption of policies restricting car use or increasing its price through taxation as we have already noted in dealing with individual countries. The construction, maintenance and servicing of cars and highways is a very profitable business and an important source of employment. Moreover, despite its harmful social and environmental effects, the automobile is unquestionably a very popular mode of transport among individual consumers. The more car-oriented a country becomes, the greater the political pressure opposing policies that make car ownership and use more expensive or more difficult. Increasing car orientation can eventually pass the point of no return, leading to complete dependence. This situation already exists in many American cities and, as we have shown, many European urban areas are heading in the same direction.

Some countries, such as the Netherlands and Germany, have attempted – with some success – to resist automobile dominance. They have already set a good example by adopting policies which promote public transport, walking and cycling, and discourage the low density suburban sprawl that gives the car a monopoly on travel. In these countries, even though the automobile remains the most important transport mode, significant and coherent measures – such as those mentioned in the previous section – have allowed a more balanced transport system, thus indicating that political will, coupled with technical measures, may indeed be successful.

Some urban transport problems (environmental degradation, for instance) are no longer of national interest only. As shown by the EU's interest in this field (see the two EC 'green' papers on the Urban Environment in 1990 and on the impact of Transport on the Environment in 1992), the desire to intervene in urban areas is great since policies must be coordinated among countries. International cooperation is necessary for manufacturers to standardize technological improvements, and thus minimize their costs. Moreover, failure to coordinate could lead to incompatibilities that diminish the benefits of such standards. For example, Germans driving cars with catalytic converters currently have difficulties finding unleaded petrol in certain southern and eastern European countries. Public transport is also the subject of EU policy. For instance, the EC directive to open competition in the public transport sector to operators from all member states may have a significant impact on cost, as shown by the more limited deregulation programmes in Scandinavia.

Whatever the advantages and disadvantages of the various policies enumerated in this book, it is clear that no one policy is sufficient for dealing with the multi-faceted urban transport problem. A tailored set of differentiated policies is needed to handle the different situation in each country, and even in different cities in the same country. However, the introduction of such policies seems unlikely. It is certain that the underpricing of car ownership and use (by the failure to force automobile users to pay the full social, environmental and economic costs) distorts modal split in favour of the automobile. The higher the car's share of modal split, the more difficult it is to get political support for such pricing measures. Moreover, the fiscal austerity in the public sector is likely to continue, and this will increase the pressure to reduce public transport subsidies, thus putting public transport at an increasing competitive disadvantage. The future looks bleak both for urban transport and for our cities: more traffic jams, more pollution and reduced accessibility. Urban transport would seem to provide yet another illustration of the difficulties inherent in governing urban areas in the year 2000.

Bibliography

Gart Ademe (1993) Cetur, *Les déplacements urbains en province: pratiques et opinions* (Paris: Cetur).

A. Altshuler, J. Womack and J. Pucher (1979) *The Urban Transportation System: Politics and Policy Innovation* (Cambridge: MIT Press).

Advisory Commission on Intergovernmental Relations (ACIR) (1984) *Fiscal Disparities: Central Cities and Suburbs* (Washington: ACIR).

B. Andersen (1993) 'Regulatory Reform in Local Public Transport in Scandinavia', in D. Banister and J. Berechman (eds) (1993) *Transport in a Unified Europe* (Amsterdam: Elsevier), pp. 249–290.

American Automobile Manufacturers Association (1993) *Automobile Facts and Figures* (Detroit: American Automobile Manufacturers Association).

American Public Transit Association (1970–1994, annual) *Transit Factbook* (Washington, D.C.: American Public Transit Association).

M. Anselmetti (1994) *Le trasformazioni socioeconomiche in Lombardia negli anni '80: l'incidenza sul settore dei trasporti* (Milano: Regione Lombardia, Servizio Reti e sistemi di Trasporto).

Associazione degli Interessi Metropolitani (AIM) (1989) *La disciplina della sosta nelle aree urbane* (Milano: AIM).

Automobile Club d'Italia (ACI) (1993) Proceedings of the 1993 Conferenza del Traffico e della circolazione (Stresa: ACI).

Azienda Trasporti Municipali di Milano (Atm) (1993) *Milano: Territorio-Mobilità, trasporto-inquinamento* (Milano: Atm).

A. Baanders (1994) 'La mise en place des régulations: l'expérience aux Pays-Bas', *Colloquium Se déplacer au quotidien dans trente ans* (Paris) (March).

D. Banister (1989) 'Urban congestion and gridlock in Britain', *Built Environment*, 15 (3), 166–75.

D. Banister and J. Berechman (1993) *Transport in a Unified Europe* (Amsterdam: Elsevier).

D. Banister, J. Berechman and G. De Rus (1992) 'Competitive Regimes within the European Bus Industry: Theory and Practice', *Transportation Research*, 26A (2), 167–78.

J.H. Bater (1980) *The Soviet City: Ideal and Reality* (Beverly Hills: Sage Publications).

D. Bayliss (1991) 'Transport in London: Entering the 1990s', *Built Environment*, 17 (2), 107–21.

P. Blair (1985) 'Metros and the Soviet City', *Soviet Geography: Review and Translation*, 26 (6) (June) 456–60.

E. Bruehning (1993) 'Traffic Safety in Eastern and Western Europe at the Beginning of the Nineties', *Transport Reviews*, 13 (3), 265–76.

K. Button and P. Rietveld (1993) 'Financing urban transport projects in Europe', *Transportation*, 20, 251–65.

Canadian Urban Transit Association (1980–1992, annual) *Summary of Canadian Transit Statistics* (Toronto: Canadian Urban Transit Association).

Canadian Urban Transit Association (1990) *The Environmental Benefits of Urban Transit* (Toronto: Canadian Urban Transit Association)

Canadian Urban Transit Association (1991) *The Implications of Demographic and Socioeconomic Trends for Urban Transit in Canada* (Toronto: Canadian Urban Transit Association).

Canadian Urban Transit Association (1992) *Canadian Transit Factbook: 1991 Operating Data for Individual Transit Systems* (Toronto: Canadian Urban Transit Association).

Canadian Urban Transit Association (1993) *Summary of Provincial Funding Programs for Transit* (Toronto: Canadian Urban Transit Association).

Censis (1986) *Indagine sulla mobilita passageri nelle grandi aree metropolitane* (Rome: Censis).

Central Statistical Office of the Soviet Union (1975, 1980, 1983, 1985, 1988) *Economic Statistics* (Moscow: State Committee for Statistics)

R. Cervero (1986a) 'Urban Transit in Canada: Integration and Innovation at its Best', *Transportation Quarterly*, 40 (3) (July), 293–316.

R. Cervero (1986b) *Suburban Gridlock* (New Brunswick: Center for Urban Policy Research, Rutgers University).

R. Cervero (1989) 'Jobs – Housing Balancing and Regional Mobility', *Journal of the American Planning Association*, 55, 136–50.

Cetur (1975–1994, annual) *Données et analyses sur les transports urbains* (Paris: Ministère des Transports).

Cetur (1988) *Les chiffres du stationnement* (Paris: Cetur).

Cetur (1990a) 'Financement des opérations de transport en milieu urbain', *Dossier Déplacements* (Paris: Cetur).

Cetur (1990b) *Ville plus sûre, quartiers sans accidents: savoir faire et techniques* (Paris: Cetur).

Cetur (1990c) 'Partenariats', *Déplacements*, 2 (Paris: Cetur).

Cetur (1992a) 'Gérer le stationnement, un métier pour un service urbain', *Dossier Déplacements Urbains* (Paris: Cetur) (January).

Cetur (1992b) *Guide zone 30: methodologie et recommandations* (Paris, Cetur).

Cetur (1992c) 'Italie', *Déplacements*, 7.

Cetur (1992d) 'Partenariats public/privé', *Déplacements*, 8.

Cetur (1993a) 'Déplacements et liens sociaux', *Déplacements*, 14.

Cetur (1993b) 'L'enjeu de la desserte du périurbain', *Déplacements*, 13.

Cetur (1994) *Les enjeux des politiques de déplacement dans une stratégie urbaine* (Paris: Cetur).

Commissariat Général du Plan (1992) *Transport 2010* (Paris: Documentation Française).

Commissariat Général du Plan (1993) *Villes, démocratie, solidarité: le pari d'une politique* (Paris: Documentation Française).

Commission of the European Communities (1990) *Green Paper on the Urban Environment* (Brussels, EEC).

Commission of the European Communities (1992) *Green Paper on the impact of Transport on the Environment* (Brussels: EEC).

CROW (1992) *Still more Bikes behind the Dikes* ('The Netherlands: Centre for Research and Contract Standardization in Civil Engineering').

S. Cullinane (1993) 'United Kingdom: Deregulated Transport in an Over-Regulated Continent', in I. Salomon *et al.* (eds), *A Billion Trips a Day* (Dordrecht: Kluwer).

M. Dasgupta (1994) 'Travel Demand and Policy Impacts: European Experiences' (Crowthorne: TRL), unpublished paper.

G. Del Sole and M. Pazienti (1993) 'Italy: a (Motorway) Bridge to the South', in I. Salomon, P. Bovy and J. P. Orfeuil, *A Billion Trips a Day* (Dordrecht: Kluwer).

DoT (1984) *Buses* (London: HMSO).

DoT (1985) *Transport Act* (London: HMSO).

DoT (1986) *Transport Statistics Great Britain: 1975–1985* (London: HMSO).

DoT (1988) *National Travel Survey: 1985/86 report* (London: HMSO).

DoT (1989) *New Roads by New Means* (London: HMSO).

DoT (1993a) *Transport Statistics for London: 1993* (London: HMSO).

DoT (1993b) *Transport Statistics Great Britain* (London: HMSO).

DoT (1993c) *National Travel Survey 1989/91* (London: HMSO)

A. Downs (1992) *Stuck in Traffic: Coping with Peak-Hour Traffic Congestion* (Washington, D.C.: Brooking Institution)

DREIF (various years) *Les transports de voyageurs en Ile-de-France* (Paris: DREIF).

A. Dumov (1985) 'A Look at Soviet Public Transport', *Passenger Transport*, 43 (1), 1–6.

Environmental Protection Agency (1993) *National Air Quality and Emission Trends Report* (Research Triangle Park, NC: Office of Air Quality Planning and Standards).

European Conference of the Ministers of Transport (ECMT) (1985) *Changing Patterns of Urban Travel* (Paris: ECMT).

R. English (1994) 'Automakers Face Tougher Fuel Emissions Rules', *The Financial Post* (February 26), 5–11.

S. Estrias and X. Richter (1993) 'Industrial Restructuring and Microeconomic Adjustment in Poland', *Comparative Economic Studies*, 35 (4) (Winter), 1–20.

European Federation for Transport and Environment (1994) *Green Transport: A Survey* (Brussels).

Eurostat (1993) *Transport: Annual Statistics, 1970–1990* (Brussels: Office for official publications of the European Communities).

Federal Transit Administration (1992a) *National Transit Summaries and Trends* (Washington, D.C.: US Department of Transportation).

Federal Transit Administration (1992b) *Federal Transit Act, As Amended through June 1992, and Related Laws* (Washington, D.C.: US Department of Transportation).

Federal Highway Administration (1970–1993, annual). *Highway Statistics* (Washington, D.C.: US Department of Transportation).

Federal Highway Administration (1992) *Nationwide Personal Transportation Survey* (Washington, D.C.: US Department of Transportation).

Federal Highway Administration (1993) *Journey-to-Work Trends in the USA and its Major Metropolitan Areas, 1960 to 1990* (Washington, D.C.: US Department of Transportation).

Fiat (1989) *Mobilità e aree urbane* (Torino: Fiat).

G. Förschner (1990) 'Entwicklung der Verkehrsmittelwahl in Städten der DDR im Lichte des Systems repräsentativer Verkehrsbefragungen (SrV) 1972 bis 1987', *Die Strasse*, 30 (11), 323–6.

G. Förschner (1992) *Mobilität und Verkehrsmittelwahl in unterschiedlichen Stadtgrössengruppen der neuen Bundesländer* (Dresden: Germany: Technische Universität Dresden).

G. Förschner and E. Schöppe (1992) 'Erste Ergebnisse des erweiterten Systems repräsentativer Verkehrsbefragungen SrV-Plus 1991', *Strassenverkehrstechnik*, 2, 84–91.

R.A. French (1987) 'Changing Spatial Patterns in Soviet Cities – Planning or Pragmatism', *Urban Geography*, 8 (4), 309–20.

G. Fromm (1992) 'Mehr Geld für mehr Projekte: Das neue Gesetz zur Verkehrsfinanzierung in den Städten' *Der Städtetag*, 5, 342–7.

Y. Geffrin and M. Muller (1993) *Evolution démographique, croissance urbaine et mobilité* (Paris: Cetur).

German Ministry of Transport (1982–1993, annual), *Verkehr in Zahlen* (Bonn, Germany: Bundesverkehrsministerium).

German Ministry of Transport (1980–1994, annual), *Finanzielle Leistungen des Bundes, der Länder und der Kommunen für den Öffentlichen Personennahverkehr* (Bonn, Germany: Bundesverkehrsministerium).

German Ministry of Transport (1988) Forschung Stadtverkehr: Kooperationen im ÖPNV (Bonn: Bundesverkehrsministerium).

German Ministry of Transport (1992), *Finanzierung des ÖPNV in den neuen Bundesländern* (Bonn: Bundesverkehrsministerium).

German Press and Information Service (1992) *Tatsachen über Deutschland* (Frankfurt am Main: Societätsverlag).

P. Gordon, H. Richardson and M. Jun (1991) 'The Commuting Paradox: Evidence from the Top Twenty', *Journal of the American Planning Association*, 57 (4) (Autumn), 416–20.

S. Grava (1984) 'Urban Transport in the Soviet Union', in H.W. Morton and R.C. Stuart (eds), *The Contemporary Soviet City* (New York: Sharpe).

S. Grava (1988) 'The Planned Metro of Riga: Is it Necessary or Even Desirable?', *Transportation Quarterly*, 43 (3), 451–72.

G. Guiliano (1992) 'An Assessment of the Political Acceptability of Congestion Pricing', *Transportation*, 19 (4), 335–58.

J.M. Guidez (1990) *10 ans de mobilité urbaine: les années 80* (Paris: Cetur).

J.G. Hajdu (1989) 'Pedestrian Malls in West Germany: Perceptions of their Role and Stages in their Development' *Journal of the American Planning Association*, 54 (3), 325–35.

P. Hall and C. Hass-Klau (1985) *Can Rail Save the City? The Impacts of Rail Rapid Transit and Pedestrianisation on British and German Cities* (London: Gower).

M. Hamer (1987) *Wheels Within Wheels: A Study of the Road Lobby* (London: Routledge & Kegan Paul).

T. Hart (1992) 'Transport, the Urban Pattern and Regional Change, 1960–2010', *Urban Studies* 29 (3), 483–503.

J. Hartman (1990) 'The Delft Bicycle Network', in R. Tolley, *The Greening of Urban Transport: Planning for Walking and Cycling in Western Cities* (London).

D. Harvey (1989) *The Urban Experience* (Baltimore: Johns Hopkins University Press).

C. Hass-Klau (1990) *The Pedestrian and City Traffic* (London: Belhaven Press).

C. Heidemann, U. Kunert and D. Zumkeller (1993) 'Germany: A Review at the Verge of a New Era', in I. Salomon, P. Bovy and J.P. Orfeuil (eds) *A Billion Trips a Day* (Dordrecht: Kluwer).

J. Heilbrun (1987) *Urban Economics* (New York: St Martin's Press).

V. Himanen, P. Nijkamp and J. Padjen (1992) 'Environmental Quality and Transport Policy in Europe', *Transportation Research*, 26–A (2), 147–57.

W. Hook (1994) 'Eastern Europe: Paving the Way to Environmental Disaster?', *Sustainable Transport*, 3 (June), 4–16.

Hungarian Ministry of Transport and Communications (1992) *The Hungarian Transport Policy* (Budapest: Ministry of Transport and Communications).

Hungarian Ministry of Transport (1993) *Transport Data, 1983–1992* (Budapest: Hungarian Ministry of Transport, Communications, and Water Management).

R. Hupfer (1992) *Die Gestaltung der Güter- und Personenbeförderung in Ungarn* (Budapest: Ministry of Transport and Communications).

B. Hutchison (1991) 'Metropolitan Transportation: Patterns and Planning', in T. Bunting and P. Filion (eds), *Canadian Cities in Transition* (Toronto: Oxford University Press).

Ifo Institut (1992) *Verkehrskonjunktur 1992: Abgeschwächtes Wachstum im Westen – beginnende Erholung im Osten* (München: Ifo Institut für Wirtschaftsforschung).

International Monetary Fund (1993) *Government Statistical Yearbook* (Washington, D.C.: IMF).

International Roadway Federation (various years) *World Road Statistics* (Washington, D.C. and Geneva: IRF).

INRETS (1989) *Un milliard de déplacements par semaine* (Paris: Documentation Française).

Institute of Transportation Engineering of Prague (1991) *Dopravni Informace Praha: 1990* (Prague: Ustav Dopravniho Inzenyrstvi Hlavniho Mesta Prahy)

Istat and Automobile Club d'Italia (1992) *Statistica degli incidenti stradali, annuario, n. 39* (Rome: Istat and Aci).

Iveco (1990) *Per un nuovo Trasporto Pubblico Urbano* (Rome: Iveco).

Iveco (1993) *Trasporto pubblico locale: due obiettivi convergenti: approvare la nuova disciplina e sostenere il rinnovo del parco* (Rome: Iveco).

G. Jansen (1992) *Commuting in Europe: Homes Sprawl, Jobs Sprawl, Traffic Problems Grow* (Delft: TNO Policy Research).

P. Jones (1991) 'UK Public Attitudes to Urban Traffic Problems and Possible Countermeasures: A Poll of Polls', *Environment and Planning C, Government and Policy*, 9, 245–56.

P. Jones (1992) 'Urban Road Pricing: Dealing with the Issue of Public Acceptability: A UK Perspective', in Raux, C. and M.Lee-Gosselin (eds), *La mobilité urbaine: de la paralysie au péage* (Grenoble: Programme Rhone-Alpes, Recherche en Sciences Humaines).

P. Jones and A. Hervik (1992) 'Restraining Car Traffic in European Cities: An Emerging Role for Road Pricing', *Transportation Research*, 26A (2), 133–45.

J. Kenworthy (1991) 'The Land Use and Transit Connection in Toronto: Some Lessons for Australian Cities', *The Australian Planner* (September), 149–54.

J. Kessler and W. Schroeer (1993) *Meeting Mobility and Air Quality Goals: Strategies that Work* (Washington, D.C.: Environmental Protection Agency).

H. Kitchen (1990) 'Transportation', in R. Loreto and T. Price (eds), *Urban Policy Issues: Canadian Perspectives* (Toronto: McClelland & Steward).

J. Klofac (1989) 'Transportation Engineering and Urban Transport in Czechoslovakia', *Journal of International Association of Traffic and Safety Sciences*, 13 (2), 43–52.

K. Klopotov (1983) 'Urban Transport in the USSR', *UITP Revue*, 32 (3) 223–8.

Koninklijk Nederland Vervoer (1981–93, annual) *Kerncijfers Personenvervoer* (The Hague: KNV).

W. Korver, G. Jansen and P. Bovy (1992) *The Netherlands: Ground Transport Below Sea Level* (Delft: TNO-Beleidsstudies).

M. Kroon (1990) 'Traffic and Environmental Policy in the Netherlands', in R. Tolley, *The Greening of Urban Transport: Planning for Walking and Cycling in Western Cities* (London).

U. Kunert (1988) 'National Policy Towards Cars in the Federal Republic of Germany', *Transport Reviews*, 8 (1), 59–74.

P. Lassave (1987) *L'expérience des plans de déplacements urbains (1983–1986)* (Paris: Cetur).

C. Lefèvre (1989) *La crise des transports publics*, Notes et Etudes Documentaires, 4900 (Paris: Documentation Française).

C. Lefèvre (1991) 'I sistemi di trasporto su ferro e la creazione di Transit authorities: alcune grandi città europee a confronto', *Territorio*, 10, 201–11.

C. Lefèvre and J.M. Offner (1990) *Les transports urbains en question: usages, décisions, territoires* (Paris: Celse).

C. Lefèvre and J.M. Offner (1993) 'Urban Transport Management in French Cities: Who Governs?', *Urban Affairs Quarterly*, 28 (3), 480–500.

L. Lesley (1989) 'Assessment of an Eastern European City's Public Transport System' *Transportation Research*, 23A, (2), 129–37.

T. Lijewski (1987) 'Transport in Warsaw', *Transport Reviews*, 7 (2), 95–118.

D. Lipton and J. Sachs (1990) 'Creating a Market Economy in Eastern Europe: The Case of Poland', *Brookings Papers on Economic Activity* (Washington, D.C.: Brookings Institution).

LITRA (1994) *Chronique des transports, 1993–94* (Bern: LITRA).

F. Loiseau (1989) *Le piéton, la sécurité routière et l'aménagement de l'espace public* (Paris: Cetur).

D. Lorrain (1991) 'Public Goods and Private Operators in France', in R. Batley and G. Stoker, *Local Government in Europe: trends and development* (London: Macmillan), 89–109.

M & T (various issues, various years).

R. Mackett (1992) 'Transport Planning and Operating in a Changing Economic and Political Environment: The Case of Hungary', *Transport Reviews*, 12 (1), 77–96.

V. Malfatti (1993) *Il trasporto pubblico locale* (Milan: Il Cardo).

H.U. Mann, R. Mueck, M. Schubert, H. Hautzinger and R. Hamacher (1991) *Personenverkehrsprognose 2010 für Deutschland* (München: Interplan Consult GmbH and Heilbronn: Institut für angewandte Verkehrs- und Tourismusforschung e.V).

H. McClintock (1990) 'Planning for the Bicyle in Urban Britain: An Assessment of Experience and Issues', in R. Tolley (ed.), *The Greening of Urban Transport: Planning for Walking and Cycling in Western Cities* (London).

Ministero dei Trasporti (1984) *I primi anni dell'attività normativa delle regioni nella logica della L.151* (Rome: Direzione Generale Programmazione, Organizzazione e Coordinamento).

Ministero dei Trasporti (1985, 1987 and 1993) *Conto Nazionale dei Trasporti*, (Rome: Direzione Generale Programmazione, Organizzazione e Coordinamento, Istituto Poligrafico e Zecca dello Stato).

Ministero dei Trasporti (1986) *Il trasporto pubblico locale: analisi per regione, anno 1984* (Rome: Direzione Generale Programmazione, Organizzazione e Coordinamento).

Ministero dei Trasporti (1993a) *Il trasporto pubblico locale* (Rome: Direzione Generale Programmazione, Organizzazione e Coordinamento).

Ministero dei Trasporti (1993b) *Le ferrovie in concessione e in gestione governativa* (Roma: Direzione Generale Programmazione, Organizzazione e Coordinamento).

Ministry of Transport, Public Works and Water Management (n.d.) *Bicycle First* (Rotterdam: Ministry of Transport).

Ministry of Transport, Public Works and Water Management (1993) *Facts about Cycling in the Netherlands* (Rotterdam: Ministry of Transport).

S. Mitric (1992) *The Warsaw Urban Transport Review* (Washington, D.C.: The World Bank).

S. Mitric (1993) 'Urban Public Transport Policies and Strategies in Eastern Europe', paper presented at the CODATU VI Conference, Tunis, Tunisia (February).

S. Mitric (1994) *The Poland Urban Transport Review* (Washington, D.C.: The World Bank).

A. Morchoine (1993) 'Transport et environnement: quels enjeux pour la ville?', *Transport Public* (March), 30–3.

C. Napoléon and J.C. Ziv (1981) *Le transport urbain: un enjeu pour les villes* (Paris: Dunod).

National Safety Council (annual) *Accident Facts*, (Washington, D.C.: National Safety Council).

W. Neenan (1972) *Political Economy of Urban Areas* (Chicago: Markham Press).

P. Newman and J. Kenworthy (1989a) 'Gasoline Consumption and Cities: A Comparison of U.S. Cities with a Global Survey', *Journal of the American Planning Association*, 55 (1), 24–37.

P. Newman and J. Kenworthy (1989b) *Cities and Automobile Dependence: An Intenational Sourcebook* (Aldershot : Gower).

Observatoire Régional des Déplacements (1993) *Mémento de statistiques 1993* (Paris: STP).

Organisation for Economic Cooperation and Development (OECD) (1988) *Cities and Transport* (Paris: OECD).

Organisation for Economic Cooperation and Development (OECD) (1990) *Environmental Policies for Cities in the 1990s* (Paris: OECD).

Organisation for Economic Cooperation and Development (OECD) (1991) *Urban Infrastructure: Finance and Management* (Paris: OECD).

Organisation for Economic Cooperation and Development (OECD) (1992) *Market and Government Failures in Environmental Management: the Case of Transport* (Paris: OECD).

Organisation for Economic Cooperation and Development (OECD) (1993 and 1994), *Energy Prices and Taxes* (Paris: International Energy Agency, OECD)

J.M. Offner (1993) 'Twenty Five Years (1967–1992) of Urban Transport Planning in France', *Planning Perspectives*, 8, 92–105.

Office of Urban Transport Economics (Izba Gospodarcja Komunikacji Miejskiej) (1992) 'Principal Problems of Public Transport in Poland', *Public Transport International*, 41 (4), 42–6.

J.P. Orfeuil (1993) 'France: A Centralized Country in between Regional and European Development', in I. Salomon, P. Bovy and J.P. Orfeuil, *A Billion Trips a Day* (Dordrecht: Kluwer Academic Publishers).

F. Pasquay and J. Monigl (1992) 'Issues Affecting Urban Public Transport in Eastern Europe', in *Stuctural Changes in Public Transport* (Brussels: International Association for Public Transport), 95–157.

K.E. Perrett, *et al.* (1989) 'The Effect of Bus Deregulation in the Metropolitan Areas', *Transport and Road Research Laboratory Research Report*, 210 (Crowthorne, TRRL).

M. Pharaoh and J. Russell (1991) 'Traffic Calming Policy and Performance', *Transport Reviews*, 62 (1), 79–105.

A. Pisarski (1992) *New Perspectives in Commuting* (Washington: US Department of Transportation).

J. Ploeger (1995) 'Transportation and urban space planning', *XXth World Road Congress* (Montreal), no page numbers.

Polish Association of Urban Public Transport (1994) *Urban Transport Statistics* (Warsaw: Association of Urban Public Transport).

Polish Central Statistical Office (annual) *Transportation Statistics Yearbook* (Warsaw: Central Statistical Office).

Prague Institute of Transportation Engineering (1994) *Dopravni Informace Praha* (Prague: Prague Institute of Transportation Engineering).

J. Pucher (1982) 'Discrimination in Mass Transit', *Journal of the American Planning Association* (Summer), 315–26.

J. Pucher, A. Markstedt and I. Hirschman (1983) 'Impacts of Subsidies on the Costs of Public Transport', *Journal of Transport Economics and Policy* (May), 155–76.

J. Pucher (1988a) 'Urban Travel Behaviour as the Outcome of Public Policy: the Example of Modal Split in Western Europe and North America', *Journal of the American Planning Association*, 54 (4), 509–20.

J. Pucher (1988b) 'Policy: The Key to Public Transport's Success: Comparing Experiences from Twelve Countries in Western Europe and North America', *UITP Revue*, 37 (1) (May), 65–91.

J. Pucher (1990a) 'Capitalism, Socialism, and Urban Transportation: Policies and Travel Behaviour in the East and West', *Journal of the American Planning Association* (Summer), 278–96.

J. Pucher (1990b), The East–West Transportation Spectrum', *Transportation Quarterly*, 44 (3) (July), 441–65.

J. Pucher and S. Clorer (1992a) 'Taming the Automobile in Germany', *Transportation Quarterly*, 46 (3) (July), 383–95.

J. Pucher (1992b) 'The Challenge to Public Transport in Central Europe', *Public Transport International*, 41 (4), 28–41.

J. Pucher (1993) 'The Transport Revolution in Central Europe', *Transportation Quarterly*, 47 (1) (January), 97–113.

J. Pucher and I. Hirschman (1993) 'Urban Public Transport in the USA', *Public Transport International*, 43 (3), 12–33, 47–55, 61–69.

J. Pucher (1994a) 'Canadian Public Transport: Recent Developments and Comparisons with the United States', *Transportation Quarterly*, 48 (1), 65–78.

J. Pucher (1994b) 'Modal Shift in Eastern Germany: Transportation Impacts of Political Change', *Transportation*, 21, 1–22.

C. Quin *et al.* (1990) *Le financement des transports collectifs urbains dans les pays développés* (Paris: Documentation Française).

P. Ranzani (1992) *Mobilità, trasporti e ambiente in Lombardia* (Milano: Franco Angeli).

M. Rataj (1988) *Public Transport and Modal Split in Urban Areas in Poland* (Warsaw, Poland: Warsaw Polytechnic Institute of Transportation), unpublished paper.

RATP (1992) *L'organisation des transports urbains: étude comparative européenne* (Paris: RATP Reseau 2000).

Regional Plan Association (1991) *The Renaissance of Rail Transit in America* (New York: Regional Plan Association).

M. Renner (1988) *Rethinking the Role of the Automobile* (Washington, D.C.: Worldwatch Institute).

M. Richter and J. Vogel (1986) 'Die Prager Metro', *Kraftverkehr*, 2, 40–2.

H. Riedel (1984) 'Die U-Bahnen der Sowjetunion', *Verkehr und Technik*, 9, 347–50.

S. Rosenbloom (1992) 'The Transportation Disadvantaged', in G. Gray and L. Hoel (eds), *Public Transportation* (Englewood Cliffs, N.J.: Prentice-Hall).

A. Ryder (1994) 'What Went Wrong With Poland's Warsaw Metro', *Transportation Quarterly*, 48 (1), 79–90.

J. Sachs and D. Lipton (1990) 'Poland's Economic Reform', *Foreign Affairs*, 43 (Summer), 47–66.

H. Saitz (1988) 'Erfurt City and Traffic: An Example of Transport Policy and Planning in the German Democratic Republic', *Transport Reviews*, 8 (1), 1–17.

I. Salomon, P. Bovy and J.P. Orfeuil (1993) *A Billion Trips a Day* (Dordrecht: Kluwer).

V. Sankov (1986) *The Soviet Transport System* (Moscow: Novosti Publishing House).

K. Schaeffer and E. Sclar (1980) *Access for All: Transportation and Urban Development* (New York: Columbia University Press).

H.U. Schmidt (1992) *Unfallentwicklung in Deutschland* (Dresden: Institut für Verkehrssicherheit, Hochschule für Verkehrswesen).

Second Chamber of the States-general (1990) *Second Transport Structure Plan*.

K. Sharman and M. Dasgupta (1993) *Urban Travel and Sustainable Development: An OECD/ECMT Study of 132 Cities* (Crowthorne: TRRL).

D.J. Shaw (1978) 'Planning Leningrad', *Geographical Review*, 68 (2), 183–200.

B. Simpson (1989) *Urban Rail Transit: An Appraisal*, Contractor report 140 (Crowthorne: TRRL).

R. Soberman (1983) 'Urban Transportation in the U.S. and Canada: a Canadian Perspective', *The Logistics and Transportation Review*, 19 (2), 99–109.

T. Spector (1992) *Les transports face au désenclavement des quartiers* (Paris: Cetur).

Statistics Canada (1979 and 1982) *Travel to Work* (Ottawa: Statistics Canada).

Statistics Canada (1993) *Household Travel* (Ottawa: Statistics Canada).
Statistics Canada (1994) *Canada Yearbook* (Ottawa: Statistics Canada).
W. Suchorzewski (1993) 'The Effects of Rising Car Travel in Warsaw', paper presented at the 'International Conference on Travel in the City: Making it Sustainable' (Dusseldorf, Germany) (7–9 June), 1–3.
M. Tessitore (1992) 'Différents aspects des transports urbains: mobilité, planification, transports publics et circulation dans des grandes villes' in Cetur (1992c), 'Italie', *Déplacements*, 7.
The Economist (1994) 'Against the Grain: A Survey of Poland', *Economist* (16 April), 1–22.
K. Thomas *et al.* (1989) 'Die Entwicklung des städtischen Personenverkehrs in der DDR', *Wissenschaftliche Zeitschrift der Technischen Universität Dresden*, 38 (4), 199–204.
R. Tolley (1990a) *The Greening of Urban Transport: Planning for Walking and Cycling in Western Cities* (London).
R. Tolley (1990b), 'A Hard Road: The Problems and Cycling in British Cities', in R. Tolley (ed.), *The Greening of Urban Transport: Planning for Walking and Cycling in Western Cities* (London).
H. Topp (1993) 'Parking Policies to Reduce Car Traffic in German Cities', *Transport Reviews*, 13 (1), 83–95.
Transport Canada (1994) 'Traffic Fatality Trends in Canada', special tabulations conducted for the author by Transport Canada, Ottawa.
Transports Destination 2002 (1992a) *L'exigence de l'Environnement*, Fifth colloquium (Nantes: Ministry of Transport).
Transports Destination 2002 (1992b) *Demandes de transport et stratégies d'infrastructures* (Paris-La Défense: Ministry of Transport).
W. Tyson (1993) *Contractual Relations between Authorities and Operators: An International Survey on the Organisation of Local and Regional Public Transport* (Brussels, UITP).
Umwelt (1992) 'Luftreinhaltung: Schadstoffminderung bei Kraftfahrzeugen', 9, 342–4.
Umwelt (1993) 'Zehn Jahre Katalysator: Deutschland gab entscheidende Impulse für die Luftreinhaltung in ganz Europa', 10, 402–3.
US Congress (1991) *The Intermodal Surface Transportation and Efficiency Act of 1991, Public Law 102–240* (December 18).
US Department of Commerce (1993) *Statistical Abstract of the USA* (Washington, D.C.: Bureau of the Census).
US Department of Transportation (1985–1992, annual) *National Urban Mass Transportation Statistics* (Washington: Urban Mass Transit Administration and Federal Transit Administration).
US Department of Transportation (1990–1993, annual) *National Transportation Statistics* (Washington, D.C.: Bureau of Transportation Statistics).
UTP (1993) Les chiffres clés du transport urbain (Paris: UTP).
UTP (1994) *Les chiffres clés du transport urbain: 1993* (Paris: UTP).
J. Vanderwagen (1992) *Transit in Canada: A Handbook for Environmentalists* (Vancouver: Greenpeace Canada).
P. Varlaki and L. Lesley (1986) 'Modal Split and Urban Public Transport Management in an Eastern European Country', *Transportation*, 13, 235–55.
Verband deutscher Verkehrsunternehmen (1991) *VDV-Statistik 1990* (Cologne: Verband deutscher Verkehrsunternehmen).

Verband deutscher Verkehrsunternehmen (1992) *OEPNV-Finanzierungskonzept vor dem Hintergrund von Bahnreform und Regionalisierung* (Cologne: Verband deutscher Verkehrsunternehmen).

Verkehrswissenschaftlicher Verein (1988) *Aufgabenteilung im Städtischen Personenverkehr* (Die 19. Budapester internationale Beratung für Verkehrsplanung und Verkehrstechnik: Budapest, Hungary).

V. Vlassov (1984), 'Standard and Social Aspects of the Development of Passenger Transport in the Cities of the USSR', *UITP Revue*, 33 (4), 297–300.

W. Voigt (1991) 'Die Verkehrssituation in den neuen Bundesländern: Probleme und Aufgaben', in *Urban Development Policy – Transport Development Policy* (Budapest: Scientific Association for Transport).

W. Voigt (1988a) 'Town Traffic in the German Democratic Republic', *Transport Reviews*, 8 (3), 183–95.

W. Voigt (1988b) 'Grundsätze der Aufgabenteilung im städtischen Personenverkehr der DDR', in Verkehrswissenschaftlicher Verein, *Aufgabenteilung im Staedtischen Personenverkehr* (Budapest), 38–47.

M. Wachs (1993) 'Learning from Los Angeles: Transport, Urban Form, and Air Quality', *Transportation*, 20, 329–54.

D. Walmsley and K. Perrett (1992) *The Effects of Rapid Transit on Public Transport and Urban Development, State of the Art Review*, 6 (Crowthorne: TRRL).

S. Warner (1972) *The Urban Wilderness: A History of the American City* (New York: Harper & Row).

F.W. Webster, P. Bly, R. Johnston, N. Pauley and M. Dasgupta (1986) 'Changing Patterns of Urban Travel', *Transport Reviews,* 6 (1), 49–86 and 6 (2), 129–72.

E. Weiner (1992) *Urban Transportation Planning in the USA* (Washington, D.C.: US Department of Transportation).

P. White (1987) *Public Transport: Its Planning, Management and Operation* (London: UCL Press).

P. White and S. Cassidy (1992) 'Local Bus Deregulation and the Consumer', *Consumer Policy*, 2 (2), 77–80.

P. White (1993) 'Road Passenger Transport and Deregulation', *Public Money and Management,* (January–March), 35–40.

World Resources Institute (1992) *The Going Rate: What it Really Costs to Drive* (Washington, D.C.: World Resources Institute).

M. Yeates (1990) *The North American City* (New York: Harper & Row).

B. Younes (1993) 'Roads in Urban Areas: To Build or not to Build?', *Transport Reviews* 13 (2), 99–116.

Index

Note: when page numbers are in **bold** type, this indicates that the entry is of general concern rather than being concerned with a specific country.